Automation, Collaboration, & E-Services

Chris Bissell and Chris Dillon (Eds.)

Ways of Thinking, Ways of Seeing

Mathematical and Other Modelling in Engineering and Technology

Springer

Editors

Dr. Chris Bissell
The Open University
Faculty of Mathematics,
Computing and Technology
Walton Hall
Milton Keynes, MK7 6AA
UK
E-mail: c.c.bissell@open.ac.uk

Dr. Chris Dillon
The Open University
Faculty of Mathematics,
Computing and Technology
Walton Hall
Milton Keynes, MK7 6AA
UK
E-mail: c.r.dillon@open.ac.uk

ISBN 978-3-642-25208-2
e-ISBN 978-3-642-25209-9

DOI 10.1007/978-3-642-25209-9

Automation, Collaboration, & E-Services
ISSN 2193-472X

Library of Congress Control Number: 2011941488

Typeset & Cover Design: Scientific Publishing Services Pvt. Ltd., Chennai, India.

Printed on acid-free paper

9 8 7 6 5 4 3 2 1

springer.com

Preface

Chris Bissell and Chris Dillon

This is a book about 'models' and 'modelling', concepts that are extraordinarily difficult to pin down. As pointed out by a number of authors of this volume, a model could be a scale model, a mathematical model, a sketch, a segment of computer code, an analogy, a working device, or many other things. A model can be used to describe or explain some aspect of the natural world, it can be used as part of the design of an artefact, it can be part of an attempt to convince someone of some argument or ideology, or to determine public or corporate policy. The objective of this book is to investigate, in both a historical and a contemporary context, such aspects of 'modelling'.

Turning first to engineering and technology, these areas, like science, often use mathematical descriptions of the world. Because engineering models use many of the same mathematical techniques as scientific models (differential equations, Fourier and Laplace transformations, vectors, tensors, for example) it is easy to assume that they are one and the same in essence. Yet in the case of engineering (and technology in general) such models are much more likely to be used to design devices, equipment or industrial plant than for representing and analysing natural objects or phenomena. This means that historically there has been just as great – if not greater – emphasis on rules-of-thumb, charts, and empirical models as there has been on analytical models (although the latter have also been vitally important in areas such as electronics, mechanics, chemical and civil engineering, and so on). Other important tools have been scale models of various types, analogue simulators, and analogue and digital computation. A characteristic of all these approaches has been to develop new ways of seeing, thinking and talking about natural and artificial objects and systems.

Much of what has been written about mathematical modelling has been:

- The casting of traditional mathematics in 'modelling' dress, but still very theoretical, such as Gershenfeld (1999).
- A detailed study of historical and/or contemporary practice in the natural sciences or economics, such as Morgan & Morrison (1999).
- Highly theoretical and specialised examinations of engineering modelling, such as Lucertini et al (2004).
- The formal use of computers in converting well-specified modelling assumptions into computer code, see, for example Kramer (2007).

This book aims to redress the balance by studying primarily modelling in technological practice (although some chapters address essentially non-technological areas where there is an important link to engineering and technology). It is worth noting that some of the models considered in this book are not always highly valued in formal education at university level, which often takes an 'applied science' approach close to that of the natural sciences (something that can result in disaffection on the part of students). Yet in an informal context, such as design departments, laboratories, industrial placements, policy making, and so on, a very different situation obtains. A number of chapters will consider such epistemological aspects, as well as the status of different types of models within practitioner communities.

Professional training in many areas often includes a great deal of mathematics, conventionally said to 'underpin' the various disciplines. Yet practitioners often claim never to have used the majority of the mathematics they were taught. If by this they mean, for example, that they rarely if ever solve differential equations, invert matrices, or use vector calculus then – unless they are working in highly specialised research and development – they are almost certainly correct. This apparent paradox is best understood by examining the social context of the use of mathematics by the vast majority of professionals. Consider, for example, information engineering – disciplines such as telecommunications, electrical/electronics engineering, control engineering and signal processing that form the focus of several chapters of this book. Information engineers have developed visual or pictorial ways of representing systems that not only avoid the use of complex mathematics (although the techniques may well be isomorphic with the conventional formulations taught in universities), but have enabled ways of seeing and talking about systems that draw on the graphical features of the models.

This approach develops within a community of engineering practice where the interpretation and understanding of these visual representations of systems behaviour are learnt, shared and become part of the normal way of talking. Engineers put models to work by using them as the focal point for a story or conversation about how a system behaves and how that behaviour can be changed. It is by mediating in this process – acting to focus language by stressing some features of the real system while ignoring others – that models contribute to new shared understandings in a community of engineering practice. Interestingly, modern computer tools continue to exploit many much earlier information engineering techniques – techniques originally designed to eliminate computation but now used primarily to facilitate communication and human-machine interaction.

Very similar arguments apply to the use of mathematics and modelling in areas outside engineering. Models deriving originally from scientific or engineering principles may have to inform policy making, or be used to develop robust computer software, and the way this is best done is not always clear or straightforward; very often it is highly contested. A number of chapters consider this aspect of modelling in the context of economics, climate change, epidemiology and software development.

Themes that are highlighted in this volume include:

- The role of language: the models developed for engineering design have resulted in new ways of talking about technological systems.
- Communities of practice: related to the previous point, particular engineering communities have particular ways of sharing and developing knowledge.
- Graphical (re)presentation: engineers have developed many ways of reducing quite complex mathematical models to more simple representations.
- Reification: highly abstract mathematical models are turned into 'objects' that can be manipulated almost like components of a physical system.
- Machines: not only the currently ubiquitous digital computer, but also older analogue devices – slide rules, physical models, wind tunnels and other small-scale simulators, as well as mechanical, electrical and electronic analogue computers.
- Mathematics and modelling as a bridging tool between disciplines.
- Modelling in large-scale socio-technological contexts, such as climate change and epidemiology.
- A move away from rigid formalism in software engineering.

The wide-ranging first chapter, by John Monk, looks at historical and philosophical aspects of modelling. Beginning with the nineteenth century, and the work of Lodge, Maxwell, Thomson, Kirchhoff, Mach and others – all predominantly physicists, but all of whom were enormously influential on electrical and mechanical technologies – Monk examines the use in particular of hydraulic and mechanical analogies and physical models. The consideration of Mach's philosophical bent then leads naturally to a study of a number of twentieth century philosophers, who have been intrigued by the notion of a model, including Wittgenstein, Foucault and Rorty. This initial chapter sets the scene for much that follows.

Chapter 2, by John Bissell, moves from the wide-ranging to the highly specific: a variety of approaches in science and technology based on the notions of dimensional analysis and dimensional reasoning. Like many of the themes of this book, the dimensional approach is one which is of enormous utility for scientific and technological practice, but one which finds only a minor place in the professional education of scientists and engineers. Bissell gives a number of examples of the power of this approach – from scale modelling in engineering design to estimating parameters of nuclear explosions from limited data.

The following two chapters, by Chris Dillon and Chris Bissell, can be usefully considered together, as both of them examine the role of language, communities of practice, and graphical modelling tools in information engineering. One important theoretical 'underpinning' of these disciplines (although the authors of these two chapters might well contest such a notion of 'underpinning') is the mathematics of complex numbers and vector calculus. From the turn of the twentieth century up to the 1950s information engineers invented a range of 'meta-mathematical' techniques to enable them to free electronics, telecommunications and control engineering design practice from the difficulties engendered by such a mathematical basis of their models. Furthermore, such tools that enabled mid-twentieth century engineers to avoid complicated calculation have now become an essential element in the human computer interface design of CAD software.

If the previous two chapters concentrated on 'meta-mathematical' tools in the forms of tables, maps, graphs and charts, then Charles Care's Chapter 5 turns to the physical devices used in such modelling before the days of the digital computer: physical models, electrical analogies and analogue computers. There are echoes of both Chapter 1 and Chapter 4 here, as Care looks in more detail at some of the work of Lord Kelvin (William Thomson) as well as analogue devices used in control engineering and related areas. But he also examines direct analogues such as soap films, electrolytic tanks and wax models, topics which have been very much under-researched in the history of modelling (Care, 2010).

Chapter 6 represents something of a turning point in the book, as it documents some of the ways that the technological modelling tradition centred on cybernetics and even control engineering – closely related to several earlier chapters – was transferred to the human domain through systems thinking (Ramage and Shipp, 2009). Magnus Ramage and Karen Shipp discuss the systems dynamics approach deriving from Jay Forrester and others; the work of Stafford Beer on organisations; Howard Odum on ecological systems; and the unique systems diagramming approach of the former Faculty of Technology at the UK Open University (now incorporated into the OU's Faculty of Mathematics, Computing and Technology.)

The remaining three chapters, in a sense, focus more on the human element – although this is not to underestimate the human factors in the book as a whole. In Chapter 7 Marcel Boumans situates Dutch computer-based economic modelling of the 1980s in the context of a history of the analogue modelling of economic systems. It is a particularly interesting case study, as the system concerned, FYSIOEN, is a *computer* model of a *hydraulic* model of an economy – that itself looks back to a famous physical hydraulic model of the 1950s (Bissell, 2007). Ultimately FYSIOEN was not particularly successful, yet there are many lessons to be learned from the attempt.

Chapter 8, by Gabriele Gramelsberger and Erika Mansnerus, turns to contemporary policy issues, looking at how models of infectious disease transmission and climate change can inform decision making. The chapter covers the philosophical framework of such modelling and contrasts the inner world (empirical knowledge, computational framework, etc) and the outer world (predictive and prognostic power) of such modelling. It also considers the 'story telling power' of models, something that also emerges in a number of other chapters.

Finally, Chapter 9, by Meurig Beynon, describes the 'Empirical Modelling' approach to software construction developed over a number of years at the University of Warwick, UK, which draws particularly on the philosophy of William James and the sociology of Bruno Latour. Essentially this is a call to move away from an excessively reductionist and algorithmic approach to software development. All three of Chapters 7 – 9 echo the ideas about models as analogues and the pragmatic philosophy of Chapter 1.

These essays will be of interest to scientists, mathematicians and engineers, but also to sociologists, historians and philosophers of science and technology. And, although one or two of the chapters include significant mathematical content, understanding the fine details of this is not necessary in order to appreciate the general thrust of the argument.

References

Bissell, C.C.: Historical perspectives – The Moniac. A Hydromechanical Analog Computer of the 1950s. IEEE Control Systems Magazine 27(1), 59–64 (2007)

Care, C.: Technology for modelling: electrical analogies, engineering practice, and the development of analogue computing. Springer, London (2010)

Gershenfeld, N.: The Nature of Mathematical Modelling. Cambridge University Press, Cambridge (1999)

Kramer, J.: Is abstraction the key to computing? Communications of the Association for Computing Machinery 50(4), 37–42 (2007)

Lucertini, M., Gasca, A.M., Nicolò, F. (eds.): Technological concepts and mathematical models in the evaluation of modern engineering systems. Birkhäuser, Basel (2004)

Models as mediators. In: Morgan, M.S., Morrison, M. (eds.) Ideas in Context, Cambridge University Press, Cambridge (1999)

Ramage, M., Shipp, K.: Systems Thinkers. Springer, London (2009)

Contents

Contributors

MEURIG BEYNON is Emeritus Reader at the University of Warwick, UK. His major research interests are in the empirical modelling area outlined in his chapter in this volume.

CHRIS BISSELL is Professor of Telematics at The Open University, UK, where he teaches many aspects of ICT and researches primarily the history of technology.

JOHN BISSELL, at the time of writing his chapter, was completing his PhD in plasma physics at Imperial College of Science, Technology and Medicine, UK.

MARCEL BOUMANS is Associate Professor in the Faculty of Business and Economics at the University of Amsterdam, Co-Director of the faculty research programme 'Methodology and History of Economics' and Co-Editor of the Journal of Economic Thought.

CHARLES CARE completed a PhD on the history of analogue simulation at the University of Warwick, UK, and has subsequently worked for British Telecom as a Senior Software Engineer while retaining an Associate Fellowship at Warwick.

CHRIS DILLON was, until his recent retirement, Director of the Centre for Outcomes-Based Education at The Open University, UK. Before then he was Senior Lecturer in Electronics at the Open University and an author of course materials in control engineering, signal processing and other areas.

GABRIELE GRAMELSBERGER is a philosopher at the Freie Universität, Berlin, who is interested in the influence of computation on science and society. She has conducted an extensive study on the influence of computer based simulation on science, in particular in climate research and cell biology.

ERIKA MANSNERUS is a Postdoctoral Research Fellow at the London School of Economics. She studies the use of computational tools in public health interventions, such as vaccination planning or pandemic preparedness work, particularly the benefits and limitations of these tools and to how to interpret the evidence produced by them.

JOHN MONK is Emeritus Professor of Digital Electronics at The Open University, where he contributed to many courses in electronics, control engineering, robotics, and related areas.

MAGUS RAMAGE is Lecturer in Systems at The Open University, UK. He has researched the history of systems thinking and is currently investigating the notion of 'information' in an interdisciplinary context.

KAREN SHIPP recently retired as Lecturer in Systems at The Open University, UK. Before this she was an educational software developer and designed numerous suites of computer aided learning for OU courses.

Chapter 1
Creating Reality

John Monk

The Open University, UK

Abstract. Analogues in the nineteenth century provided experimenters such as Lodge, Maxwell, Kirchhoff, Mach and Hertz with inspiration for mechanical descriptions of hidden physical processes that had, for example, electrical or magnetic properties, and suggested mechanical models that could illustrate their developing theories to a wider audience. Models and theories intertwine, since any confirmation or test of a theory has to show its predictive power in a specific situation. It is tempting to imagine that a model or theory is an accurate reflection of what takes place in reality; however prominent nineteenth century physicists and latterly pragmatist philosophers have insisted that our descriptions of reality are of our own making and are a product of our institutions and customs. Models as part of our descriptive practices, therefore, make a contribution to the construction of reality. This chapter discusses some of the nineteenth-century analogical models that offered ways of seeing and understanding physical phenomena, and goes on to discuss how philosophers have explored ways of thinking about the relationship between models and reality.

1.1 Modelling as a Discursive Practice

In an article entitled *Models*, Ludwig Boltzmann (1902) described a model as a 'representation … of an object'. He noted that 'real objects' can be modelled in thought, and that 'objects in thought' can be represented by writing in a notebook. Kühne (2005), answering the question 'What is a model?' unpromisingly pointed out, 'there is little consensus about what exactly a model is' but he also claimed that 'All … models are linguistic in nature' and that in software engineering, for example, a model is 'formulated in a modelling language' which employs diagrammatic and graphical notations[1]. Similarly Kulakowski et al. (2007, 54) regarded system models as being articulated as 'verbal text, plots and graphs, tables of relevant numerical data, or mathematical equations'. In a theological paper that, most unusually, made reference to the physical sciences, Cosgrave (1983) saw

[1] See Beynon Chapter 9 in this volume for further discussion of software engineering models.

C. Bissell and C. Dillon (Eds.): Ways of Thinking, Ways of Seeing, ACES 1, pp. 1–28.
springerlink.com

models as images, metaphors, analogies or symbols that are familiar and 'taken from ordinary life' which, for theology at least, intermingle with references to 'particular religious or theological objects or realities', an approach that has resonances with some of the philosophical perspectives discussed later in this chapter. Hesse (1953) likewise observed that 'most physicists do not regard models as literal descriptions of nature, but as standing in a relation of analogy to nature'. The emphasis on metaphor and analogy suggests that models and modelling have strong connections with other kinds of imaginative descriptive practices. Indeed Frigg (2010) argued that modelling has strong similarities with fiction writing and Vorms (2011) declared that '[m]odels such as the simple pendulum, isolated populations, and perfectly rational agents, play a central role in theorizing' and are 'imaginary ...fictional or abstract entities'.

It is tempting to imagine that a model or theory is an accurate reflection of what takes place in reality; however, prominent nineteenth century physicists and latterly pragmatist philosophers have insisted that our descriptions of reality are of our own making and are a product of our institutions and customs. Models as part of our descriptive practices, therefore, make a contribution to the construction of reality. This chapter discusses some of the nineteenth-century analogical models that offered ways of seeing and understanding physical phenomena, and goes on to discuss how philosophers such as Peirce, Wittgenstein, Foucault, Fleck and Rorty have explored ways of thinking about the relationship between models and reality.

1.2 Ways of Seeing: Some Nineteenth-Century Analogues

For nineteenth century physicists the term 'model' often referred primarily to mechanical constructions. Boltzmann (1892), for example, described an exhibition of mathematical models, many of which were made from 'plaster casts, models with fixed and movable strings, links, and all kinds of joints' but he also took the opportunity to mention 'mechanical fictions' – models or analogies that were 'dynamical illustrations in the fancy' that aided the development of mathematical statements of theories without necessarily ever being built. While theories were intended to be universal, any demonstration of a theory or hypothesis demanded descriptions of particular situations. The textbooks of the time described numerous examples of simple situations, for instance, of a single swinging pendulum (Maxwell 1876a, 100). Such descriptions of situations or artefacts used to illustrate a physical theory might today be called models.

The emergence of theories about electromagnetic phenomena provided a rich environment for the development of modelling practices (Hesse 1953). Phenomena such as sparks, electric shocks and forces between electromagnets and pieces of iron, attributed to electricity and magnetism, hint at out-of-sight physical activity. A demonstration may show what happens in a particular situation but does not comprehensively expose the relationships between such phenomena. How, then, are these relationships to be explained? One option is to speculate about the form of microscopic, or otherwise imperceptible, processes that bind phenomena together; another is to circumvent any explanation and systematically record observations about what happens and the conditions under which events occur, and then try to summarize a way of calculating what will happen from knowledge of the prevalent conditions; a third option is to construct an analogue – a mechanism that is not intended to represent the hidden physical workings but has visible or visualizable parts that, in some way, bear a parallel relationship to observable electrical phenomena.

Analogues in the nineteenth century provided physicists such as Lodge, Maxwell, Kirchhoff, Mach and Hertz with inspiration for mechanical descriptions of hidden physical processes that resulted in, for example, observable electrical or magnetic phenomena, and suggested mechanical models that could illustrate their developing theories to a wider audience. Models and theories intertwine, since any confirmation or illustration of a theory has to show its predictive power in a specific situation, while avoiding misleading or over-elaborate explanations that serve to obfuscate rather then illuminate the phenomena. The advantage of having a plausible mechanism in mind is that it breaks down the relationships between phenomena into components. If these correspond to familiar mechanical components in well-defined configurations, then the model helps to suggest and visualize relationships between them. The mechanism, or plan, or diagram of a mechanism, then acts as a systematic record and a familiar reminder and, by nature of the implied constraints on the physical behaviour of the mechanism, restricts the grammar of accounts of what can happen.

1.2.1 Oliver Lodge

In his book entitled *Modern Views on Electricity* and published in 1889, Oliver Lodge aimed to 'explain without technicalities … the position of thinkers on electrical subjects' and he chose to do this with 'mechanical models and analogies' (Lodge 1889, v). In the text he avoided mathematics, but in a handful of places and in the appendix he told something about contemporary developments 'in less popular language than in the body of the book' (Lodge 1889, 387). His primary audience, however, was those who had 'some difficulty' with the published theories (Lodge 1889, v).

To avoid ignorance of the state of knowledge of electricity, one option, Lodge proposed, was to accept an analogy. He anticipated that his readers would 'get a more real grasp of the subject and insight into the actual processes occurring in Nature' by becoming familiar with analogies (Lodge 1889, 61), and he advised that the alternative was to use mathematics and dispense with 'pictorial images' (Lodge 1889, 13). Although some mathematicians rejected 'mental imagery', it was nevertheless helpful to have 'some mental picture' alongside the 'hard and rigid mathematical equations' (Lodge 1889, 61).

A promotional page at the front of Lodge's book was explicit about the text developing the 'incompressible-fluid' idea of electricity" (Lodge 1889, v), and in the subsequent text Lodge suggested that electricity 'behaves like a perfect and all-permeating liquid' and 'obeys the same laws' (Lodge 1889, 12). But he warned his readers against becoming too attached to the analogy and inferring that 'because electricity obeys the laws of a liquid therefore it is one' (Lodge 1889 p.12). He considered the possibility that electricity and fluids may be 'really identical' to be a 'fancy', and counseled vigilance for discrepancies between the behavior of the two that would undermine their apparent similarity.

An important electrical device at the time – and still a vital part of electronics, but now predominantly at the microscopic level – was the condenser, or Leyden jar (now known as a capacitor), used to store electrical charge. Lodge exploited several analogues to explain the operation of Leyden jars. These devices were made of glass and had separate conducting coatings on their inside and outside. Connections were made to the interior and exterior coatings of a jar, which could retain an electric charge. One analogue used elastic, cords and beads. Another was a hydraulic model built from parts that could be bought from a plumbers' shop. As well as including an etching of the hydraulic model Lodge's book showed a 'skeleton diagram', reproduced here in Figure 1.1, and provided instructions for making the model. These began by telling the reader to 'procure a thin india-rubber bag, such as are distended with gas at toy-shops'. The bag, or balloon, was set inside a globular glass flask and the whole apparatus filled with water (Lodge 1889, 54–55).

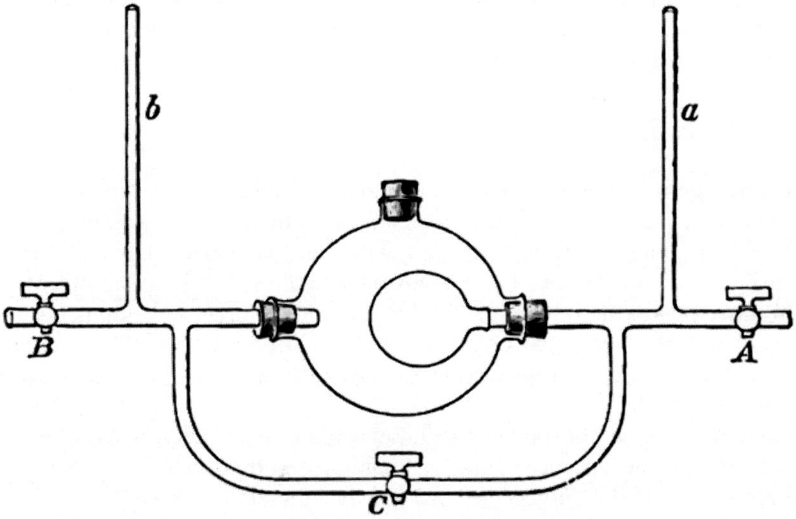

Fig. 1.1 Lodge's hydraulic analogue of a Leyden jar. Source: Lodge (1889, 55).

When stopcock *C* was closed the rubber bag or balloon could be inflated with water by a pump connected through the open stopcock at *A*. The water displaced by the distending bag flowed out to a tank through the open stopcock *B*. In the analogue the water pump represents a generator of electricity, the inside of the inflatable rubber bag corresponds to the inner conducting surface of the Leyden jar, and the outer surface of the bag corresponds to the outer conducting surface of the jar. Having established an analogical connection with an electrical device, the behaviour of the hydraulic model could be translated speculatively into electrical activity. On their own, the diagrams, pictures and descriptions of the construction of the hydraulic model did not indicate how the analogue might be operated; nevertheless, the diagrams, pictures and prose imposed constraints on any narrative of the model's operation, and offered ways of thinking about the Leyden jar's behaviour.

Lodge himself described a number of possible hydraulic experiments that had parallels with demonstrations carried out on a Leyden jar. For instance, the hydraulic pump connected to the open stopcock *A* could be turned on and the rise in pressure indicated by a rise in the height of water in the narrow vertical tube, *a*. When the stopcock *B* was open to a tank of water 'the pump can be steadily worked, so as to distend the bag and raise the gauge *a* to its full height, *b* remaining at zero all the time'. If stop-cock *A* was then closed the balloon would remain inflated unless the valve *C* was opened when the stretched balloon would force water out via *C* and suck water in to the globular glass flask through the opening next to *B* (Lodge 1889, 56).

When Lodge described the discharge he wrote, 'by the use of the discharger *C* the fluid can be transferred' and, although he was describing the experiment on the hydraulic model, he continued his sentence using terms that would normally be applied to a Leyden jar and adds 'from inner to outer coat'. He then reverted to terms appropriate to the apparatus constructed from the water-filled rubber bag to conclude 'the strain relieved, and the gauges equalized'. In Lodge's account the two sides of the analogy become entangled and the vocabulary and grammar of one domain is used in the other.

The Leyden jar, Lodge openly declared, has a 'hydrostatic analogue', and parts of the hydraulic equipment could be identified with parts of the Leyden jar. It would be easy to see, and it is easy to imagine, the rubber bag inflating and deflating and, when inflated, exerting pressure on the water so as to force water out on one side and draw it in on the other. The apparatus is unnecessary; a picture and a few words are sufficient for readers acquainted with balloons.

The electrical mechanism of the Leyden jar is inaccessible to human senses, nonetheless the hydraulic model offers figures of speech that can be a part of a plausible explanation of the electrical phenomena. However, caution is needed for not all electrical phenomena will have parallels in the field of hydraulics. For example, Lodge listed a number of effects of electrical currents and voltages that occurred outside of the wires carrying them. One such is the effect that an electric current has on a magnetic compass, which cannot be portrayed by any analogous effect occurring outside pipework carrying water (Lodge 1889, 91–92).

Limited by the hydraulic analogy Lodge introduced a new model for electromagnetism and electromagnetic fields. Fig. 1.2 is one of his diagrams. The rack is supposed to be an analogue of an electrical conductor with the linear movement of the rack mirroring the movement of electricity in a wire. In the analogue, the rack engages with an array of cogs that rotate when the rack moves, with the speed of rotation of the cogs standing for the strength of the magnetic field. All the cogs with plus signs on one side of the rack would rotate in the same direction, and all those with plus signs on the other side would rotate in the opposite direction; those cogs with minus signs made the mechanism plausible but unfortunately for the analogy would rotate in the opposite direction to those with the plus signs.

Fig. 1.2 Lodge's model of an electromagnetic field using a rack and meshing cogs. Source: Lodge (1889, 186).

To overcome the difficulty Lodge suggested the cogs with minus signs were related to 'negative electricity' (Lodge 1889, 264).

Lodge exploited this analogue of rotating wheels to explain 'electric inertia', or 'self-induction' (Lodge 1889, 186–187). He viewed the cogs as flywheels, so that to accelerate the enmeshed cogs some effort is necessary to move the rack. If the effort is no longer applied the 'motion is prolonged for a short time by the inertia'. In materials such as iron the magnetic effect is pronounced, and to emulate such substances, Lodge declared, they 'have their ... wheel-work exceedingly massive'. Lodge added various modifications to the model to refine the analogy, such as introducing wheels that can slip, leading him to write 'A magnetized medium ... is thus to be regarded as full of spinning wheels ... imperfectly cogged together'. Lodge's mechanical analogue illustrated a radical theory of electromagnetism that was, at the same time, a simple model of a straight, current-carrying wire.

Electrical activity is often invisible but, through various analogies, Lodge created fantastical explanations that could predict electrical effects in a convincing way. His choices of mechanical and hydraulic analogues were plausible although they did not validate the use of the analogy as an explanation of electrical phenomena. Instead the analogues served as useful and effective rhetorical devices. To help his readers understand and visualise what was going on Lodge suggested that they should think of:

> electrical phenomena as produced by an all-permeating liquid embedded in a jelly; think of conductors as holes and pipes in this jelly, of an electrical machine as a pump, of charge as excess or defect, of attraction as due to strain, of discharge as bursting, of the discharge of a Leyden jar as a springing back or recoil, oscillating till its energy has gone. (Lodge 1889, 61–62).

Such fantasies can become so remote from personal experience that they cease to be persuasive or, as Lodge pointed out in the case of the hydraulic analogy for the Leyden jar, they risk being pressed too far. In some instances, however, analogies can offer an acceptable way of accommodating and debating hidden relationships between observable phenomena.

1.2.2 James Clerk Maxwell

Lodge was not the first to consider rotational elements in the explanation of electromagnetic phenomena. Helmholtz (1867), for example, observed 'a remarkable analogy between the vortex-motion of fluids and the electro-magnetic action of electric currents'. William Thomson (1872) endorsed 'Helmholtz's exquisite theory of vortex-motion' especially because the analogy showed how a theory might be developed – a challenge that Maxwell (1873a, 416) took up by looking at electromagnetic fields from a mechanical viewpoint and supposing that they were 'occupied with innumerable vortices of revolving matter' (Maxwell 1862). But he 'found great difficulty in conceiving of the existence of vortices in a medium, side by side, revolving in the same direction' (Maxwell 1861, 469) until, like Lodge after him, he likened the relationship between the flow of electric current and magnetic lines of force 'to the relation of a toothed wheel or rack to wheels which it drives (Maxwell 1861).

Lodge clearly drew on Maxwell's analogy but offered a slightly different interpretation. Lodge initially described an analogy that included wheels connected by rubber bands and then introduced Maxwell's approach reporting: "It consisted of a series of massive wheels, connected together ... by a row of elastic particles or 'idle wheels'" (Lodge 1889, 263). Maxwell, like Lodge, had introduced two sets of wheels and explained: "when two wheels are intended to revolve in the same direction, a wheel is placed between them so as to be in gear with both, and this wheel is called an 'idle wheel'". But unlike Lodge, Maxwell, whose focus was on a vortex model and its mathematical expression, avoided mentioning negative electricity by not placing an interpretation on the 'idle wheels', writing:

> The hypothesis about the vortices which I have to suggest is that a layer of particles, acting as idle wheels, is interposed between each vortex and the next, so that each vortex has a tendency to make the neighbouring vortices revolve in the same direction with itself. (Maxwell 1861)

In Maxwell's case, therefore, the layer of counter-rotating wheels were included to keep the analogy intact but the idle wheels played no part in the electrical interpretation. Lodge's mention of negative electricity did not invalidate his descriptions; however, it added an additional term to the theory that Maxwell had eliminated. In his attempt to follow the mechanical analogue closely Lodge had complicated the theory. As Hesse (1953) observed, redundant elements can also crop up in mathematical models. Referring to fragments of mathematics that might form a model she noted that a mathematical model is:

> not an isolated collection of equations ..., but is a recognisable part of the whole structure of abstract mathematics, and this is true whether the symbols employed have any concrete physical interpretation or not. (Hesse, 1953)

In doing so she revealed that some mathematical fragments in a model may not be related to anything observable, but are included to ensure the model fits an established mathematical form. To maintain a consistent description these instrumental elements of the model could be treated as references to hypothetical objects or quantities. Without a physical counterpart, however, the new object or quantity has to remain metaphysical, and is vulnerable to a reformulation of the model that might render the hypothetical object or quantity redundant.

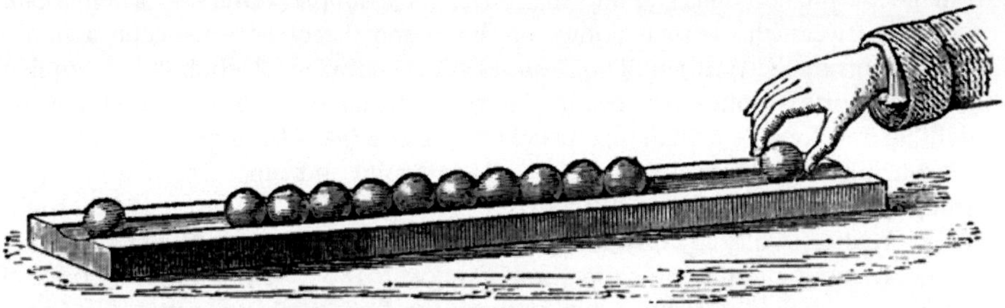

Fig. 1.3 Tyndall's analogue for sound propagation. Source: Tyndall (1867, 3).

In John Tyndall's book, Maxwell (1872, vi) admiringly noted, 'the doctrines of science are forcibly impressed on the mind by well-chosen illustrative experiments'. He was similarly complimentary about Tyndall's metaphoric imagery which illustrated the inscrutability of the physical world as 'a sanctuary of minuteness and power where molecules [build] up in secret the forms of visible things' (Maxwell 1890). In a controversial address, Tyndall (1874) explained that the senses are all we have to experience the world; all we can do is make inferences and treat what we sense as 'symbols'.

Accordingly he saw the outcome of a science education as an ability 'to picture with the eye of the mind those operations which entirely elude the eye of the body' and used analogy to encourage this. For example, he used a 'row of glass balls', illustrated in Fig. 1.3 as an image for explaining the propagation of the sound produced by a bursting balloon. Nudging the first ball would cause a movement which propagated along the row until the 'last ball only of the row flies away'. He anticipated this analogy would create an image of sound propagating from 'particle to particle through the air' (Tyndall 1867, 3–5).

Maxwell used imagery effectively to prompt questions. He suggested 'One kind of motion of the æther is evidently a wave-motion'. Deploying an analogy, he asked 'How will such waves affect an atom? Will they propel it forward like the driftwood which is flung upon the shore, or will they draw it back like the shingle which is carried out by the returning wave?' (Maxwell 1873b). What was needed, Maxwell (1864) emphasized, was 'a clear physical conception' in the analogue. In outlining the connection between calculus and mechanics he proposed a well defined mechanism 'free from friction, destitute of inertia, and incapable of being strained by the action of the applied forces'. This mechanism was not to be built but was 'to assist the imagination' and provide an alternative to algebra (Maxwell 1873a, 185).

There was, however, a psychological difficulty which Maxwell illustrated with 'the analogy between the phenomena of self-induction and those of the motion of material bodies'. He warned that it becomes difficult to abandon the mechanical analogy, or recognize that it is misleading, because our familiarity with the movement of material objects 'is so interwoven with our forms of thought that, when ever we catch a glimpse of it ..., we feel that a path is before us leading ... to the complete understanding' (Maxwell 1873a, 181). Consequently he was concerned about assumptions that are 'not warranted by experimental evidence' and cautioned about concluding, for instance, 'the electric current is really a current of a material substance, or a double current, or whether its velocity is great or small' (Maxwell 1873a, 202).

However, Maxwell (1873a, 201) applauded the 'many analogies between the electric current and a current of a material fluid'. Faraday's speculations on why an interrupted electric current should give an electric shock 'when we consider one particular wire only', Maxwell concluded, brought to bear phenomena 'exactly analogous to those of a pipe full of water flowing in a continued stream' (Maxwell 1873a, 180) and with pride announced 'the analogy between statical electricity and fluid motion turns out more perfect than we might have supposed' (Maxwell 1864).

But Maxwell's interests were far wider than the study of electricity; in his work on colour he assumed colours can be 'represented in quantity and quality by the magnitude and direction of straight lines' then, he concluded, 'the rule for the composition of colours is identical with that for the composition of forces in mechanics' (Maxwell 1860). He took the 'diffusion of heat from a part of the medium initially hotter or colder than the rest' as an analogy for the 'diffusion and decay' of an electric current induced in one circuit by a current in another. The result was that a calculation involving forces was transformed into a calculation involving heat (Maxwell 1873a, 397–398).

In the field of properties of materials Maxwell (1878) made the observation that a twisted wire 'creeps back towards its original position', and that such a wire exposed to twisting first one way and then another will also exhibit creep. He illustrated what happens with a series of different analogies: one involved the fall in temperature of 'a very large ball of iron' exposed to a series of temperature changes; another referred to the decay of electrical potential in a Leyden jar that was repeatedly charged and discharged; and the final illustration employed the decline of magnetism in iron and steel after a succession of changes in magnetisation.

Analogies transpose words and phrases into new settings. Maxwell suggested 'Scientific Metaphor' was a suitable phrase for describing a figure of speech that is transferred from 'the language … of a familiar science to one with which we are less acquainted' (Maxwell 1890). When studying moving bodies and profiting from the resources of the mathematicians, Maxwell encouraged the retranslation 'from the language of the calculus into the language of dynamics, so that our words may call up the mental image, … of some property of moving bodies … intelligible without the use of symbols.' (Maxwell 1873a, 185–194) If the language is to be scientific, Maxwell wrote, 'each term in its metaphorical use retains all the formal relations to the other terms of the system which it had in its original use' and, for example, should help those familiar with dynamics to become acquainted with electrical theories (Maxwell 1890).

In commenting on tackling the incompleteness of theories about electricity Maxwell (1864) considered that an early step was to provide 'simplification and reduction of the results of previous investigation'. He illustrated one kind of simplification with the identification of a theoretical particle which he described as 'A body so small that, for the purposes of our investigation, the distances between its different parts may be neglected', although the term particle may be applied to a planet, or even the Sun, when 'the actions of different parts of these bodies does not come under our notice'. On the other hand 'Even an atom, when we consider it as capable of rotation, must be regarded as consisting of many material particles'. (Maxwell 1876a, 11–12).

One form of simplification in a 'scientific procedure [involves] … marking out a certain region or subject as the field of our investigations' and then, Maxwell proposed, ignoring 'the rest of the universe'. In a physical science, therefore, we identify a physical system 'which we make the subject of our statements'. This system can be as simple or as complex as we choose, and involve just a few particles or bodies, or the entire material universe (Maxwell 1876a, 10). Such a simplification is evident in construction of diagrams where 'no attempt is made to represent those features of the actual material system which are not the special object of our study' (Maxwell 1876b).

Abstractions and simplifications have profound implications. A generalization can allow, for example, a single model to relate to a number of different objects by stressing particular characteristics while ignoring details of individual peculiarities and weaker associations. Further, abstraction suggests considering something apart from any specific material embodiment or conventional ways of thinking. Thus when Rubenstein (1974, 192) wrote 'A model is an abstract description of the real world' he implied that the material form of the model was not necessarily significant; the model could have a role in deliberations in a number of conceivable realities.

Another approach is to provide an analogy which contains an object, or system, and a grossly simplified environment; for example, a pendulum suspended under constant gravity (the system) within an environment that provides only the initial push to start the pendulum swinging. Interactions are then divided into two sorts: the inputs, such as the push, which originate outside the system but affect its behaviour, and the outputs, such as the pendulum's resulting motion, which are generated by the system (Kulakowski et al. 2007, 1–2). This simplification introduces causation; the inputs cause the model's behaviour and the outputs are the resulting behaviour. As with any simplification, however, the results can be misleading if the model or analogy is expected to give a good account of behaviour when the simplifying assumptions about the system or the environment do not obtain.

Maxwell was a profligate user of analogies. His strategy was to seek understanding of a collection of relationships in a poorly formulated field by adopting an analogy from another familiar and thoroughly explored domain. This was an intermediate step towards introducing and adapting a mathematical formulation which was, therefore, analogical. A set of phenomena might be explained by one or more different analogies; hence a range of analogies offers alternative ways of thinking about constructing a theory.

1.2.3 William Thomson (Lord Kelvin)

William Thomson (1824–1907), later Lord Kelvin (Anon 1892), seldom used the word 'model' in his technical papers. However, he often illustrated his popular lectures with mechanical models and he famously suggested that a measure of understanding in physics is 'Can we make a mechanical model of it?' (Thomson 1910, 830). What Thomson was seeking was an explanation of puzzling phenomena in terms of familiar objects. He wished to understand the phenomena of light, for example, 'without introducing things that we understand even less of' and the lack of a familiar analogue was why he believed he could not grasp electromagnetics (Thompson 1910, 835–836).

Thomson saw models, diagrams and examples as ways of illuminating the workings of the physical world. He expressed 'admiration for Maxwell's mechanical model of electromagnetic induction' because it was 'immensely instructive' and would assist the development of electromagnetic theory. And, in commenting on an insistence on the existence of a fixed relation between compressibility and rigidity (which is evidently flouted by a material such as jelly) Thomson berated Laplace, Lagrange and Poisson for their 'vicious habit ... of not using examples and diagrams'. For Thomson, models were pedagogical but did not necessarily represent reality; for example, although he viewed a mechanical illustration of 'the molecular constitution of solids' as 'undoubtedly instructive' the model was 'not to be accepted as true in nature' (Thompson 1910, 830).

Occasionally, however, he sought structures to satisfy his presumption that '[w]e cannot suppose all dead matter to be without form and void, and without any structure' and, he insisted, the use of a model was not 'merely playing at theory' but suggested possibilities for how molecules might be arranged (Thomson 1889).

In a lecture entitled *The Size of Atoms*, Thomson demonstrated wave propagation using a physical model made with wood and wire. This apparatus was also depicted in a full page woodcut in the report of the lecture, together with a footnote giving details of its construction. It was apparently made of a 'series of equal and similar bars ... of which the ends represent molecules of the medium'. The dimensions of the components and their materials were given and additional constructional details were noted, such as the way each wooden bar was attached to the supporting wire (Thomson 1894a).

There is no doubt in this case, therefore, that when Thomson referred to the model he was referring to the physical apparatus. Although the constructional details were relevant primarily for someone wanting to replicate Thomson's apparatus, it was the demonstrations showing the behaviour of the model that provided the visualisations of propagation. The woodcut was a satisfactory alternative for a reader who was not present at the lecture but who could imagine the model's behaviour and integrate the image with the transcript of the lecture. Through its associations with common components such as lengths of wood and wire, the material model (or its image) conjured up constraints on what might happen, as well as offering familiar terms to describe the microscopic wave phenomenon that Thomson addressed.

However, Thomson's references to models did not always indicate a model that was to be built. In his IEE President's address on the ether, he showed his audience a 'skeleton model' which was a gyrostat[2] mounted on a square frame. Next he described a web of similar 'rigid squares with their neighbouring corners joined by endless flexible inextensible threads'. The impracticality of showing the infinite network led him to ask his audience to 'imagine mounted in each one of the rigid squares of this web a gyrostat'. This loosely coupled web of framed gyrostats, he claimed, was a model of an incompressible fluid (Thomson 1889). Similarly, in speculating on the propagation of light through a substance Thomson proposed 'a model with all needful accuracy' and asked his audience to 'suppose particles of real matter arranged in the cubic order, and six steel wire spiral springs, or elastic india-rubber bands, to be hooked on to each particle and stretched between it and its six nearest neighbours'. Although this description appears to be part of a plan for a practical construction, his proposition became openly fantastical when, to eliminate gravitational effects, he suggested transporting 'the theatre of the Royal Institution … to the centre of the Earth' so he could show 'a model of an elastic solid' with a wave propagating through it (Thomson 1894a).

In contrast, some of Thomson's proposed analogues were plausible but impractical because they would be time consuming or expensive to construct. Nevertheless they were intended to stir the imagination and introduce the metaphors implied by the analogue. Thomson offered a story to explain the 'benefit' of 'electro-magnetic induction' which he compares to the 'benefit that mass is to a body shoved along against a viscous resistance'. His bizarre (but arguably plausible) allegory begins: 'Suppose, for instance, you had a railway carriage travelling through a viscous fluid'. He then switched his reference from a railway carriage to that of a boat on wheels in a viscous liquid and continued,

> We will shove off two boats with a certain velocity … but let one of them be loaded to ten times the mass of the other: it will take greater force to give it its impulse, but it will go further …

and then, to relate the story to electrical phenomena, Thomson explained:

> … [i]t requires more electric force to produce a certain amount of current, but the current goes further. (Thomson 1889).

[2] A gyrostat is a form of gyroscope.

By calling upon likely personal observations of what happens with loaded boats, Thomson's analogy offered a way of understanding what might happen in a particular electrical circuit arrangement without resorting to mathematical notation or calculation.

Although Thomson produced mechanical models to explain his theories of hidden physical phenomena, he also exploited descriptions of models that could never be built. His physical analogies were often partial but they matched his theoretical presumptions, bringing with them a descriptive vocabulary and evoking behaviours which were commonly experienced and readily accepted. Familiar but otherwise disconnected phenomena, such as electricity and heat, were coupled by imaginative combinations and situations to commonly accepted constraints of familiar mechanical components, to create an imagery that suggested satisfying explanations.

1.2.4 Gustav Kirchhoff and Ernst Mach

In a book of Lectures about Mechanics, Kirchhoff (1897, v–vi) admitted that he presented an unconventional view of mechanics. He did not believe that any theory should set out to determine causes and he aimed 'to describe, not to explain, the world of phenomena' (Boltzmann 1902). He suggested that the notion of *force* was probably introduced into theories to provide a convenient way of introducing causes, but it bred confusion and complicated the formulation of mechanics. Kirchhoff set simplicity as a criterion for a theory and claimed that all a theory of mechanics or a mechanical model required were notions of space, time and matter. Force was redundant. He conceded, however, that later developments might muster new, simpler expressions than his (Kirchhoff 1897, 1).

Mach (1911, 10) saw Kirchhoff's pronouncement that 'the problem of mechanics to be the complete and simplest description of motions' as a 'ray of hope' but reckoned that his own similar contribution predated Kirchhoff's and was 'more radical' (Mach 1919, 556). Mach challenged people who treated the theoretical entities of mass and force as a part of 'a reality beyond and independent of thought', and supported his position allegorically, writing:

> A person who knew the world only through the theatre, if brought behind the scenes and permitted to view the mechanism of the stage's action, might possibly believe that the real world also was in need of a machine-room, and that if this were once thoroughly explored, we should know all. Similarly, we, too, should beware lest the intellectual machinery, employed in the representation of the world on the stage of thought, be regarded as the basis of the real world (Mach 1919, 505).

In seeking a footing for a theory Mach (1911, 57) concluded 'One fundamental fact is not at all more intelligible than another', and therefore 'the choice of fundamental facts is a matter of convenience, history, and custom.' He recognized that an investigation tries to find relationships between appearances of things, and anything we might call a representation is just a reminder or formula 'whose form, because it is arbitrary and irrelevant, varies very easily with the standpoint of our culture' (Mach 1911, 49). Mach thought, therefore, that different forms of a theory of mechanics depended on the individuals who constructed them. Hence theories about the same object could be different (Mach 1919, 254) and, in particular, he agreed with Kirchhoff that alternative sciences of mechanics could be constructed that rendered 'the concept of force … superfluous' (Mach 1919, 255). The 'essence of science', Mach (1919, 6) claimed, was '[e]conomy of communication and of apprehension' and this was evident in the most highly developed sciences which gave 'the completest possible presentment of facts with the least possible expenditure of thought' (Mach 1919, 490).

Rhetorically, criteria provide expressions of approval rather than foundations for reasoned contributions to decisions. Thus it seems likely that Mach was registering approval when he identified 'economy of thought' with '[t]he science of physics', '[t]he mechanics of Lagrange' and 'ideas of conservation' (Mach 1919, 489; 467; 504). He was therefore sympathetic towards Kirchhoff's project when, notably omitting a reference to force, he coupled, 'the science of mechanics, in which we deal exclusively with spaces, times and masses' with those 'sciences most highly developed economically' (Mach 1919, 486).

An 'economical tendency', Mach considered, leads invariably to abstractions so that 'we never reproduce the facts in full, but only that side of them which is important to us' (Mach 1919, 482). In proposing alternatives to Newton's formulation of mechanics, Mach required 'simplicity and parsimony' to satisfy the 'economico-scientific grounds' of his project (Mach 1919, 244). He looked for labour-saving in the description of phenomena and their relationships, which drove him to seek 'methods of describing the greatest possible number of different objects at once and in the concisest manner' (Mach 1919, 6–7). Even language, for Mach, was an 'economical contrivance' for symbolizing experience (Mach 1919, 481). Another source of economy, Mach proposed, was the use of mathematics which he described as 'the economy of counting' (Mach 1919, 486), and he illustrated the economy of algebra with an analogy of a merchant who does not handle his goods but instead 'operates with bills of lading' (Mach 1919, 488).

Although Mach adopted David Hume's view of causality (Shanks 2011) he did not dismiss the use of cause and effect as part of a provisional investigation when cause and effect can be regarded as 'things of thought, having an economical office' (Mach 1919, 485). However, for 'any profound or exact investigation' he rejected the asymmetry in a causal relationship between phenomena, and regarded phenomena as 'dependent on one another in the same way that the geometer regards the sides and angles of a triangle as dependent on one another' (Mach 1919, 579).

Mach frequently asked his readers to create a picture: for instance, of 'a mass M… joined by some elastic connection with a mass m'' (Mach 1919, 202); 'a body moving vertically upwards with a definite velocity' (Mach 1919, 312); 'a ray of light' (Mach 1919, 374); 'a liquid mass confined between two similar and similarly situated surfaces very near each other' (Mach 1919 p.392); and so on. It becomes apparent that the translator, in many instances, could have substituted the word 'model' for 'picture'. Indeed, Janik (2001) in writing about the work of Mach and Hertz, uses the word model as a translation for the German *Bild* in Mach's and Hertz's work and Visser (1999), in writing about Hertz, noted the translation of the German word *Bild* is commonly 'picture', but 'physicists gradually began identifying it with what in English we would call analogy, theory, model' – an observation that makes Wittgenstein's work (see below) of greater interest for the study of models.

1.2.5 Heinrich Hertz

Heinrich Hertz, who studied physics at the University of Berlin when both Helmholtz and Kirchhoff were teaching there, declared that he owed 'very much to Mach's splendid book' (Hertz 1899, xxiv). Hertz's view was that all physics is based on the 'laws of mechanics'. He praised Maxwell for his 'discovery of mechanical analogies' (Boltzmann 1892) after he had produced a concise version of Maxwell's theory of electrodynamics. Hertz's aim was to remove 'all unessential ideas' and reduce relations between the 'essential ideas…to their simplest form', and was able to do so because Maxwell's technique had retained, in the theory's formulation, 'a number of superfluous, and in a sense rudimentary, ideas' (Hertz 1893).

Hertz's own book on Mechanics began with the proposition that finding out about the world enables us to anticipate 'future events, so that we may arrange our present affairs'. He then set out a procedure which begins with forming 'images or symbols of external objects' so the inferences from these pictured, or modelled, objects match what will happen (Hertz 1899, 1). He was at pains to note that the conceptions we form align with nature only in that the outcomes match. No other correspondences are required, and a range of models could provide a match with the appearances of an object or situation (Hertz 1899, 9). Forces, for example, are likened to '*leergehende Nebenräder*' by Hertz (1899, 14) and Mach (1908, 281; 1919, 550) – a phrase which might be translated as an idling cogs meshed with the machinery of the theory. But unlike Maxwell's and Lodge's idling wheels, which propagated motion, Hertz's metaphorical wheels played no part in the machine's functioning (Janik 2001, 157). The concept of force, Hertz considered, did not have a role at the heart of his theory of mechanics.

Hertz investigated three different formulations of mechanics: one based on Newton's work, which Hertz contested by undermining various uses of the word 'force', as in, for example, the notion of 'centrifugal force' (Hertz 1899, 6). In Hertz's second approach the 'idea of force retires in favour of the idea of energy' (Hertz 1899, 14), but this led to complicated statements, and Hertz abandoned the approach because it failed to meet his criterion of appropriateness (see below). The third formulation began, following Kirchhoff, 'with three independent fundamental conceptions, namely, those of time, space, and mass'[3] (Hertz 1899, 24–25), which Hertz hoped would give his readers a clearer picture of mechanical principles 'from which the ideas of force and the other fundamental ideas of mechanics appear stripped of the last remnant of obscurity' (Hertz 1899, xxii). Hertz's axioms were supposed to result in a simple theory and remove force as a fundamental concept but the axioms led to 'things … beyond the limits of our

[3] See Bissell Chapter 2 in this volume for a detailed discussion of dimensional analysis and dimensional reasoning.

senses' and instead of employing 'ideas of force and energy' he introduced hidden motions and masses (Hertz 1899, 26). Mach considered Hertz's third theory to be 'simpler and more beautiful, but for practical purposes our present system of mechanics is preferable' (Mach 1919, 535). Notably, however, Hertz had demonstrated that theories and models can have alternative forms.

Hertz set three criteria for judging a theory or model: correctness, permissibility and appropriateness. For Hertz a permissible model was consistent with 'the laws of thought', and correct when it led to accurate predictions (Hertz 1899, 3). Permissibility is a product of our view of logic which is likely to be invariant, but correctness depends upon the experience that institutes a model. Therefore, correctness may be nullified when new experience is gained (Hertz 1899, 3). Additionally, a model must be appropriate; the more appropriate of two models is the one with the 'smaller number of superfluous relations'. Mach observed that Hertz's criterion of appropriateness coincided with his own 'criterion of economy' (Mach 1919, 549). Appropriateness is tied up with the 'notations, definitions [and] abbreviations' deployed in a model, and depends on the purposes for which the theorizing is being done (Hertz 1899, 8). Different formulations of a theory should generate similar results but a rearrangement might also make calculating for a specific purpose more or less convenient.

For Hertz the 'crucial feature about models of physical reality … [was] that we construct them' (Janik 2001, 153). Any redundant features could be eliminated because 'our requirement of simplicity does not apply to nature, but to the images … which we fashion' (Hertz 1899, 24). Redundant features in models of physical reality, therefore, can be a product of the 'mode of portrayal' (Hertz 1899, 2). Hertz did not produce radically different theories, but rearranged old ones by challenging the fixedness of principles. He considered that corollaries can replace principles and principles can become corollaries. This perspective permitted Hertz to alter 'the choice of propositions' that he took to be fundamental and create 'various representations of the principles of mechanics' to obtain different models of things which he could then compare (Hertz 1899, 3–4).

1.3 Ways of Thinking: Some Twentieth-Century Perspectives

Perhaps the earliest proponent of the pragmatic method was Charles Sanders Peirce, a contributor to the translation of Mach's Science of Mechanics (Mach 1919, vii). Peirce's pragmatic method was a way of settling differences in metaphysical disputes by looking at the practical consequences of the two sides of the argument (James 1907, 45–47). Disputes often arise, according to Peirce (1878) when '[i]maginary distinctions are … drawn between beliefs which differ only in their mode of expression' and can be settled by simply changing how things are formulated.

Pragmatists moved philosophy in the direction of language use. Peirce (1878) attacked as self-contradictory an unidentified book on mechanics that stated 'we understand precisely the effect of force, but what force itself is we do not understand!' arguing that it was a quibble about language use and about whether we should say 'a force is an acceleration, or … it causes an acceleration'. Peirce's view on reality was that reality causes sensations that in turn cause beliefs, and that the 'essence of belief is the establishment of a habit'. In other words, beliefs lead to actions. However, a belief can also be stimulated by a fiction; in this view a model is a fiction which shapes our beliefs about reality. But, as Wittgenstein was to propose, a model can be a way of doing calculations and making predictions that is associated with techniques and practices that are recognized by appropriately trained people. Models can be viewed, therefore, as descriptions or symbols standing for a calculus for making predictions.

1.3.1 Ludwig Wittgenstein

Ludwig Wittgenstein's work afforded a connection between the nineteenth century physicists and the twentieth century philosophical pragmatists. Like Hertz before him Wittgenstein began engineering studies in Berlin. Hertz soon turned to physics (Hertz 1896, x) and Wittgenstein, who would have most likely known Hertz's text as a student, turned eventually to philosophy. Wittgenstein acknowledged the influence of both Boltzmann and Hertz on his philosophical projects (Wittgenstein 1998, 16e); he was 'a lifelong reader of Hertz' and gave his students the introduction to Hertz's *Principles of Mechanics* as an example of how to do philosophy (Janik 2001, 149). A number of scholars have supported the claim that Wittgenstein's texts provided a conduit for the philosophical remarks of Mach and Hertz (Visser 1982; Visser 1999; Janik 2001; Kjaergaard 2002).

Descriptions, Wittgenstein considered, are 'instruments for particular uses' and he gave examples of a variety of descriptive notations – 'a machine-drawing, a cross-section, an elevation with measurements, which an engineer has before him' as well as word-pictures – but he suggested we underrate their utility because we associate them with pictures that hang on the wall that merely 'portray how a thing looks' (Wittgenstein 1992, I, §291). He considered that a 'picture is ... like an illustration to a story' and 'only when one knows the story does one know the significance of the picture' (Wittgenstein 1992, I, §663), implying that a picture, or model, is situated within a practice and its significance becomes clear only when we are familiar with the practice.

In his writing Wittgenstein occasionally illustrated his argument with references to engineered artifacts. He exemplified the explanatory power of a notation, for example, by talking about a demonstration of a clock's mechanism that rotated the hour-hand when the minute-hand was turned. For the sake of argument he supposed that the demonstration did not convince the onlookers about the precise relationship between the rotations of the two hands, and that the observers were reluctant to make a prediction about the effect of further movement of the minute hand. However, they might be convinced about the relationship between the positions of the hands by seeing the arrangement of gears inside the clock, even if the gears did not move. The mechanism then becomes 'a symbol for a certain kind of behaviour'. Wittgenstein concluded 'We use a machine, or the drawing of a machine, to symbolize a particular action of the machine' but he also asked 'do we forget the possibility of their bending, breaking off, melting, and so on?' and replied unequivocally 'Yes.' (Wittgenstein 1976, 193–196; 1967, III, §33).

We therefore take references to mechanisms in two ways: one is to refer to a piece of machinery that could possibly break, the other is to use mechanism to refer to a symbol that signals the availability of operations that can be performed on the symbol itself – and part of what it symbolizes is a calculus for making predictions.

Wittgenstein wondered, 'Why is it, when designing, engineers calculate the thickness of the walls of a boiler?' Likening the design options to the choice of whether to put one's hand in a fire or not, he continued 'We shall say: human beings do in fact think' and 'this, for instance, is how they proceed when they make a boiler', then asking 'can't a boiler produced in this way explode?' and replying 'Oh, yes'. He explained that 'there are fewer boiler explosions than formerly, now that we no longer go by feeling in deciding the thickness of the walls, but make such-and-such calculations instead'. Wittgenstein's highly pragmatic conclusion was that 'we do sometimes think because it has been found to pay' (Wittgenstein 1992, §§466–470). The implication is simply that having a model – a symbol of the situation at hand – and thoughtfully applying the calculus associated with it can reduce the hazards we face. This pragmatism led Wittgenstein (1976, 86) to set usefulness as the criterion for identifying a good analogy.

Applying a calculus – calculating – is a technique and everyone performing the calculation should get the same result. It involves manipulating symbols in particular and agreed ways, and therefore demands the execution of a procedure which has to be learned. However, the acquisition of any technique also teaches an enduring way of looking at things (Wittgenstein 1967 III §35). Thus learning to use a particular model instills a corresponding kind of perceptiveness. Wittgenstein asked what was needed to promote a specific way of regarding a picture, or model. Using paintings as an analogy he responded negatively, noting that there were some pictures that did not convey anything to him, probably because 'custom and upbringing' were involved (Wittgenstein 1992, IIxi, 201e), hinting that relevant training and engagement with a practice are essential ingredients.

Reality is not a mechanism since we have neither designed it, constructed it, nor do we have faith in reality behaving as we anticipate (Wittgenstein 1967, II, §35 and §§66–69). Conversely, a calculation is not an experiment on reality. We expect computers to calculate, but we would not treat the calculation as an experiment since we pretend that we are in control. The same consideration applies to prototypes and physical analogues such as Lodge's hydraulic model of a Leyden jar, which effectively performed his audience's calculations for them. Indeed, one perspective on analogues is that they are in some way computationally equivalent to the situation to be modelled.

1.3.2 Michel Foucault

At the opening of the second lecture in his book *Modern Views of Electricity*, Oliver Lodge said that he hoped his audience was not misled by thinking he was 'going to discourse on chemistry and the latest anaesthetic' and that they realized that the word 'ether' in the title of his lecture 'means the ether, ... the hypothetical medium which is supposed to fill otherwise empty space' (Lodge 1889, 327).

Lodge's clarification presented the ether as a distinct object to be spoken about, and implicitly placed a bound on the statements he subsequently made in his lecture.

Lodge was identifying the existence of two distinct discourses: one about the ether and another about chemistry and anaesthetics. In the twentieth century Michel Foucault (1994) explored the idea of a discourse as a 'theme or a theory' which is identifiable because it organizes concepts, groups objects, and is presented in characteristic ways that restrain the style of description, the form of reasoning and the attribution of causality (Foucault 1994, 53). Nevertheless within a single discourse there can be '[q]ualitative descriptions, biographical accounts, ... deduction, statistical calculations, experimental verifications' and also 'reasonings by analogy' (Foucault 1994, 50).

As Lodge's remark suggested, there are distinctive discursive domains composed of objects and relations that each discourse constructs, and which might include 'material objects [with] ... observable physical properties' or 'fictitous objects... with arbitrary properties' that may be unchanging and consistent yet be without any other evident manifestation (Foucault 1994, 91). Discourses create objects, such as models and realities which may be situated in the past, present or future, and which necessarily rely on speculation and imagination. These objects are both created by and, in turn, shape and inform discursive practices such as conversations, drawings, writings and techniques. For example, models of electronic components, what is said about them and what techniques are used to manipulate them in designs, both spring up from and sustain a discourse about electronic engineering.

In whatever way they are present the constituents of a discourse form statements, and the coherence of a discourse is exposed by regularities amongst those statements. Such regularities may be indicated by, for instance, correlations between statements, the positioning of statements and the relationships between their functions, or through transformations of one statement into another (Foucault 1994, 27–33). A statement is always bounded by other statements and makes reference to them 'by repeating them, modifying them, or adapting them, or by opposing them, or by commenting on them' (Foucault 1994, 97–98).

In Foucault's terms a model is formed by a collection of related statements. Through its relations with other statements, the collection that shapes a model places constraints on what is said or written within the discourse, including later statements about the model. The model is thus a hybrid evolving text setting out what future statements can be made. Lodge's model of the ether, therefore, was part of an evolving discourse on the theme of electromagnetism that extended beyond his lecture or his book.

Common forms of statements are grammatical sentences and propositions (Foucault 1994, 81–84), but the verbless title of Lodge's third lecture 'The Discharge of the Leyden Jar' (Lodge 1889, 359) – which is neither propositional nor grammatical – is also a statement (Foucault 1994, 101). Statements can be 'made up of fragments of sentences, series or tables of signs, a set of propositions or equivalent formulations' (Foucault 1994, 106). Therefore, models can be composed of statements using a variety of notations. Taking Lodge's accounts of electromagnetics as an example, and given the latitude Foucault allowed, the statements creating a model include written words, diagrams and pictures, algebraic expressions, experiments and demonstrations, and the words spoken to lecture audiences. Statements acquire a status, for example, by being constantly referenced in other statements. Alternatively a statement may lose its status and be rarely mentioned, or treated as an aside, in the ensuing discourse (Foucault 1994, 99). Later statements alter the status and the relationships between earlier statements as the discourse unfolds. Consequently models evolve and may flourish, or vanish, from circulation.

Any statement, and therefore any statement about a model, has a materiality which affects its durability and accessibility. Through their materiality models can be preserved in various forms, such as in databases, museums, books, video recordings or papers (Foucault 1994, 123). The fact of a statement's materiality situates its production at a particular time and in a particular place (Foucault 1994, 101). Unlike a word or a symbol, which can be repeated, a statement is a particular instance and cannot recur. Although two statements may resemble one another, they are not identical since their unique materialities – the particular instances of their production – are part of their individual identities.

But if this is the case then how are we to generalize our models so that they become useful in different contexts and at different times? Foucault (1994, 101) recognized that by 'neutralizing' the time and place of a statement's production it can be transformed into a general form, such as a sentence spoken by different people, or the statements of Ohm's law or Newton's laws in different engineering textbooks.

Aside from the time and place of production, other types of recurrence can be neutralized – a process that deliberately or casually generates an economical description of what seems most distinctively characteristic of a range of statements. Language, for instance, is 'a system for constructing possible statements', abstracted from collections of previously enunciated statements (Foucault 1994, 85) which authorizes an unlimited number of repetitions (Foucault 1994, 27). Similarly a succession of models, each constituted from a collection of statements, may have resemblances that can be generalized to provide a specialized grammar, vocabulary or algebra that augments a discursive practice and regulates the construction of future models. Qualitative models and discursive descriptions, for example, can provide a starting point from which quantitative or algorithmic models can be built.

As in many accounts of models, Miller and Starr (1969, 145) declared that a 'model is a representation of reality'. However Foucault (1994, 48) dismisses the notion of a discourse as representing something else; statements are not signs referring to something else, but are the result of a practice that creates both a model and a reality within a single hybrid discourse. Representation becomes a relation composed of cross-references between statements. Reality is commonly placed on one side of the representing relation but that reality may be a desired future, a design or a dream, a deliberate attempt at absurdity or an image of a past reality (Foucault 1994, 90).

The model, through analogical cross-references, bridges potential gaps in a discourse about a reality by making available the discursive practices that characterise the model. For example, in a strange inversion, the exploitation of an analogy between discourses of economics – the 'reality' – and hydraulics led Phillips[4] to insert a hydraulic model into a discourse about the UK national economy that gave students a satisfying visual explanation of invisible economic processes (Leeson 2000). Similarly the analogies that Lodge deployed within the discourse on electromagnetism consisted of 'statements that concern … and belong to quite different domains of objects, and belong to quite different types of discourse' and 'serve as analogical confirmation, or … as a general principle … or … models' (Foucault 1994, 58). More generally, the formation of an analogy with elements of mathematics has created models constructed from graphical and algebraic notations which now populate the discourse of mechanics and electrical engineering.

Seen in this light, analogy is a strategy for augmenting a discourse by imitating the discursive practices of another discourse. The analogy does not have to extend to those areas of reality that either, are of no interest, are already covered by an adequate set of statements, or are better furnished by a more satisfying analogy. Use of analogy does not imply the wholesale import of practices. Statements about models, therefore, do not have to substitute for a whole discourse, and their application might consequently be seen as a simplification of a reality. The validity of a model is proven by demonstrating a coherence between a discourse providing an analogy or model, and a discourse accommodating statements accounting for personal experiences and, potentially, measurements. Measurements have the characteristics of statements because they shape what is said or written in subsequent statements, and so gain a status and a place in discourses about reality. From this perspective, measuring is an element of a discursive practice.

The presentation of statements about measurement is characteristic of the instrument used to make the measurement. A digital voltmeter, for example, provides a numeral, a pressure gauge might use a pointer, and balance scales indicate with a selection of standard weights. The author of such statements selects and uses the appropriate instrument, according to established procedures, to create a statement about current conditions. For example, to measure temperature we would select a thermometer. As Mach (1914, 349) put it 'we regulate our thoughts concerning thermal processes not according to the sensation of warmth which bodies yield us, but … by simply noting the height of the mercury'. The instrument readings are not the reality but part of a discourse about measurements. It might be argued that any set of measurements is itself a model and an analogue, filling a gap in a discourse about an otherwise ineffable reality.

A discursive practice is not an individual activity but a collectively exercised body of rules situated in a particular period, location and social, economic and linguistic arena (Foucault 1994, 117), and it is in such institutional settings that models emerge, develop and are sustained. Discursive practices derive their legitimacy by being associated with sites such as technical libraries, gateway-protected journal depositories, design offices, lecture halls, back offices, laboratories, legislative

[4] See Boumans Chapter 7 in this volume for a discussion of the Phillips hydraulic machine and the visualisation of complex economic systems.

chambers and so on. Such institutional settings spawn overlays of authority that regulate who contributes to and controls access to publications, instruments or training (Foucault 1994, 68).

The attribution of authority to engage in a discourse divides people into groups, such as the professions. In return, the authority and capacity to contribute to a discourse provides the interlocutors with an identity, evidence of competence, prestige and a warranty for the validity of their assertions (Foucault 1994, 50). Thus, it is the institutional backdrop – the combination of who is participating in the discourse and how they articulate their presentation, their education, where they are presenting and their access to instruments – that distinguishes one discourse from another (Foucault 1994, 53) and determines what constitutes acceptable models and realities.

1.3.3 Ludwik Fleck

Ludwik Fleck (1981) introduced the phrase *thought collectives* (*Denkkollektive*) to name communities that carry 'the historical development' of a 'field of thought' (Fleck 1981, 39). Although Fleck used the language of thought and cognition, his thesis can be translated into the language of Foucault by treating a thought collective as a community that maintains a discourse which, notably, institutionalizes statements that constitute models.

For Fleck (1981, 99) discourses carried with them a *thought style* manifest in a 'technical and literary style' and characteristic valuations. His thought styles establish a 'readiness for one particular way of seeing and acting' (Fleck 1981, 64). For the discourse and the models associated with a particular project or professional group, this implies that some statements are embodied and are evident in the identity of the contributor to the discourse. Thought styles also inhabit what is *not* stated. As a community grows, the discourse of the community – including its evolving models – can become a sign of solidarity. In turn, solidarity may imply that the community is resistant to development and is bound by formality, regulation and custom with a tendency to override creativity with a requirement for 'practical performance' (Fleck 1981 p.103). Models become standardised and part of the institutional canon. Electrical engineering, with its references to voltages, currents, switches, busbars, generators, transformers and so on provides an example of a large, established institution.

Different models or realities, purportedly identical, may be part of different discourses and hence objects with different attributes (Fleck 1981, 38). Similarities in vocabularies may be deceptive. Fleck (1981, 109) gave examples of individual words with different usages within different discourses: the use of the terms *force* and *energy*, for example, in the discourses of physicists, philologists or athletes, have different implications. He also warned of the antagonism that may be evident within one discourse in its assessment of statements from another. Foreign models and realities from outside the dominant discourse might be declared to be 'missing the point', the terms and statements defining models treated as irrelevant, inaccurate or in error, and any problems they aroused regarded as inconsequential.

With less exaggerated differences and in favourable circumstances, however, intersecting discourses can give rise to a flowering of metaphor and analogy that benefits both (Fleck 1981, 109).

A collection of statements 'consisting of many details and relations', as Fleck (1981 p.27) noted 'offers enduring resistance to anything that contradicts it'. Nominating a statement as a *factual* statement is a 'signal of resistance'. Usefully the mutual constraints between the statements that form the network of fact stabilize the discourse, but can also create an impression of a 'fixed reality' (Fleck 1981, 101–102). Some generalized schematic models, for example, are part of the engineering canon and treated as statements of fact that are powerful constraints on discourse. An example is the Shockley (1956/1998) model of a semiconductor diode stated as a mathematical equation which is 'widely used' (Wu et al. 2005) in discourses about electronic devices:

$$I = [\exp (qV/kT) - 1].\, I_s$$

This equation is a statement primarily relating the current I flowing through the diode and the voltage V existing across the diode. A discourse about a design referring to this model will also have to reference statements about Boltzmann's constant k, the charge on an electron q, and a characteristic of the particular diode I_s, all of which are likely to be regarded as facts. Additionally, a statement might be provided about the temperature T at which the diode is operating which, in the absence of any information about the application of the particular diode, is likely to be an assumed value.

Fleck presumed that entry into any field involved an 'apprenticeship' or induction that subjected the novice to 'authoritarian suggestion' and 'gentle constraint'. The apprenticeship develops a way of seeing that renders visible that which to those outside the discourse is invisible (Fleck 1981, 10), and is effectively an initiation into the ways of seeing, thinking and talking of a particular community of practice. Since models fill gaps in accounts of realities, an aim of any induction process is to introduce the student to the realities and adopted models of a discourse. Crucially, a rigorous induction requires experience and participation in the discourse, otherwise the 'inexperienced individual merely learns but does not discern' (Fleck 1981, 96). Levels of experience create 'a graded hierarchy of initiates' stretching from an inner, or 'esoteric' circle, to an outer, or 'exoteric' circle that is recognizable in the mode of discourse of the participants (Fleck 1981, 105).

Fleck divided the literature of the inner circle, and hence its statements about models, into three categories. These are: firstly, academic papers offering speculative models that signal a predisposition towards a fact; secondly, handbooks or reference books which originated from 'discussions among the experts, through mutual agreement … misunderstanding, … concessions and … incitement to obstinacy' (Fleck 1981, 120) and provides what is taken to be fixed and proven by the institution; and finally, there are textbooks that support the institution's traditions of teaching and whose content draws heavily on the handbooks (Fleck 1981, 112).

The models deployed by the outer circle omit small details and avoid statements that challenge the elite discourse. An example is the translation of a model of the atom described by J. J. Thomson (1904) set out for the elite in detailed technical and mathematical statements, but presented for a wider audience as 'a plum pudding with the positive electricity dispersed like currents in a dough of negative electricity' (Cockcroft 1953). The presentation is 'artistically attractive, lively and readable', points of view are treated as being either acceptable or unacceptable (Fleck 1981, 112), and facts are presented as 'immediate perceptible object[s] of reality' (Fleck 1981, 125). The models of the inner esoteric circle are often composed of cryptic, dogmatic statements ill-suited for the outer exoteric discourse. However, the elite is not isolated from popular discourse, and this exposure can force the inner circle to popularize their esoteric models in order to bolster the standing of their institutional discursive practice.

1.3.4 Richard Rorty

An account of one of William Thomson's lectures ended with the sentence, 'The discourse was illustrated by a series of experiments' (Thomson 1894b). These illustrative demonstrations were integrated into Thomson's talk and fulfilled the role of statements alongside prose, formulas, graphs, tables and physical models, all of which Richard Rorty (1991, 84–85) might have described as *texts* – a term he used to refer to all those things that people have a strong hand in, such as speaking, writing or Thomson's demonstrations.

Rorty distinguished texts from 'lumps' which, like texts, have a 'sensory appearance and spatio-temporal location' but, unlike texts, are things that people have not shaped and cannot fully control. Briefly, texts are made whereas lumps are found, and people have to be prepared to modify their descriptions of a lump when its characteristics or properties are not what they expect. Rorty centred his theories on beliefs rather than discourse, and he considered that the effect of an encounter with a text or a lump is either to reinforce beliefs or to cause beliefs to change. Uncontrollable lumpish traits are likely to cause new and surprising beliefs about reality (Rorty 1991, 97). Similarly, texts can evoke statements within a discourse about an author, or indirectly about the author's experience of texts and lumps. Once a lump has been identified, or remarked upon in some way, it becomes an object in a discourse. A lump is a part of a discourse about a reality; a model is primarily a text about the lump. Both reflect people's beliefs about the world, what it contains and how it works.

Rorty described real objects – those that have lumpish traits – as causally independent of ourselves, and he suggested that there is no need to go further and say 'our descriptions *represent* objects.' (Rorty 1991, 101). However, the verb to *represent* is in widespread use. To accommodate Rorty's position and yet retain the term *represent* we can adapt his advice on the use of the term *about* (Rorty 1991, 97). Thus when model users say 'A model represents an object' they are indicating that some statements *about the reality* are justified by statements *about a model*. For their part, model makers would indicate by the same sentence that some statements *about a model* are justified by statements *about the reality*.

Sellars (1963, 56, §50) makes a similar point about signification and his remarks could also be adapted to say that fragments of discourse setting out an object and its model are ‘two expressions, at least one of which is in *our own* vocabulary, [and which] have the same use’.

Since our links with the rest of the world are sporadic, and an omniscient view is beyond reach, the accuracy of any statement contributing towards the reality constructed by a discourse can seldom be authenticated, and any such authentication is intensely personal and transient. Modelling then becomes an activity that aims to make statements coherent rather than finding representations for objects (Rorty 1991, 106), and truth becomes an indicator of the coherency of statements. Rorty concludes that picking out which statements about an object are to be used, or trying to shoehorn a discourse into the constraints of a privileged model, are pointless tasks. Any set of statements that is ‘handy for the purposes at hand’ can be brought to bear ‘without worrying about which is closer to reality’ (Rorty 1991, 156).

1.5 Concluding Remarks

For scientists, an objective is to construct a discourse about an enduring reality; for investigators such as historians, an objective is to generate a discourse about a past reality; for interventionists, such as engineers, an objective is to construct discourses about desirable realities and future realities. In their various ways, and in their various institutions, they all maintain – and are constrained by – the institutionalized discursive practices that construct the realities they deal with.

Introducing a model may not receive institutional support. Acceptance may be achieved in two ways: firstly, by using models (and by implication a discursive practice) from an exoteric, or outer, discourse that is familiar to institutional members, although it may be alien to their institution; and secondly, by demonstrating that the model is useful to the tasks that the institution needs to undertake.

Models can also have a pedagogical role. For example, a discourse on black holes that was previously inaccessible to ‘non-experts … without the benefit of mathematics’ had barriers removed by introducing into the exoteric discourse a ‘river model of black holes’ picturing ‘space … flowing like a river into the … black hole … [and] photons, as fishes swimming fiercely in the current’ (Hamilton and Lisle 2008). The aim was to integrate explanations of current theoretical predictions into vernacular discourse.

Analogy is a strategy for modifying a discursive practice by importing fragments of discursive practice from another institution. For example, importing ideas and principles from a discourse about mechanics into an investigation of electromagnetism. The discourse remains a discourse about a reality but the import permits novel statements to be made that retain their alien identity but invigorate the host discourse. The alien statements in some instances construct a single object, and such an object is referred to as a model. The statements of a model are simply symbols for fragments of an alien calculus – the rules for a discursive practice – that both regulates and suggests how a discourse might unfold.

Once adopted, a model becomes part of the discourse about a reality that may be a fiction, but which is linked to the world outside the discourse by statements made in response to personal experiences, or by statements generated by instruments, such as measurements. A model, therefore, is embedded in a discourse and associated with those techniques and conventions for producing explanations that are acceptable within that discourse or community of practice. The assertion that a model represents an object can be translated to say 'The statements shaping the model also shape the object'. Discussions about a model, that is to say, are often starting points for new ways of thinking and talking about reality.

References

Anon: Whitehall. London Gazette, 26260, 991 (February 23, 1892)

Boltzmann, L.: On the Methods of Theoretical Physics. Proceedings of the Physical Society of London 12(1), 336–345 (1892)

Boltzmann, L.: Models. In: Wallace, D.M., Chisholm, H., Hadley, A.T. (eds.) New Volumes of the Encyclopedia Britannica, pp. 788–791. Adam and Charles Black, London (1902)

Cockcroft, J.: The Rutherford Memorial Lecture. Proceedings of the Royal Society of London. Series A 217(1128), 1–8 (1953)

Cosgrave, W.: Models of the Christian Moral Life. The Furrow 34(9), 560–574 (1983)

Fleck, L.: Genesis and Development of a Scientific Fact. Trenn, T.J., Merton, R.K. (Trans.). University of Chicago Press, Chicago (1981)

Foucault, M.: The Archaeology of Knowledge. Routledge, London (1994)

Frigg, R.: Models and fiction. Synthese 172(2), 251–268 (2010)

Hamilton, A.J.S., Lisle, J.P.: The river model of black holes. American Journal of Physics 76(6), 519–532 (2008)

von Helmholtz, H.: LXIII. On Integrals of the Hydrodynamical Equations, which Express Vortex Motion. Philosophical Magazine Series 4 33(226), 485–512 (1867)

Hertz, H.: On the Fundamental Equations of Electromagnetics for Bodies at Rest. In: Jones, D.E. (Trans.) Electric Waves, pp. 195–240. Macmillan and Co., London (1893)

Hertz, H.: Miscellaneous Papers. Jones, D.E., Schott, G.A. (Trans.). Macmillan and Co., London (1896)

Hertz, H.: The Principles of Mechanics Presented in a New Form. Jones, D.E., Walley, J.T. (Trans.). Macmillan and Co., London (1899)

Hesse, M.B.: Models in physics. The British Journal for the Philosophy of Science 4(15), 198–214 (1953)

James, W.: Pragmatism, a new name for some old ways of thinking: popular lectures on philosophy. Longmans, Green, London (1907)

Janik, A.S.: Wittgenstein's Vienna Revisited. Transaction Publishers, New Brunswick (2001)

Kirchhoff, G.: Vorlesungen über Mechanik. B. G. Teubner, Leipzig (1897)

Kjaergaard, P.C.: Hertz and Wittgenstein's Philosophy of Science. Journal for General Philosophy of Science 33(1), 121–149 (2002)

Kulakowski, B., Gardner, J., Shearer, J.: Dynamic Modelling and Control of Engineering Systems. Cambridge University Press (2007)

Kühne, T.: What is a Model? In: Bezivin, J., Heckel, R. (eds.) Dagstuhl Seminar Proceedings: Language Engineering for Model-Driven Software Development. Dagstuhl, Saarbrücken (2005)

Leeson, R.: A. W. H. Phillips: Collected Works in Contemporary Perspective, pp. 31–129. Cambridge University Press, Cambridge (2000)
Lodge, O.J.: Modern Views of Electricity. Macmillan and Co., London (1889)
Mach, E.: Die Mechanik in ihrer Entwickelung. F. A. Brockhaus, Leipzig (1908)
Mach, E.: History and Root of the Principle of the Conservation of Energy. Jourdain, P.E.B. (Trans.). Open Court Publishing Co., Chicago (1911)
Mach, E.: Contributions to the Analysis of the Sensations. Williams, C.M. (Trans.). Open Court Publishing Co., Chicago (1914)
Mach, E.: The Science of Mechanics. McCormack, T.J. (Trans.). The Open Court Publishing Co., London (1919)
Maxwell, J.C.: On the Theory of Compound Colours, and the Relations of the Colours of the Spectrum. Philosophical Transactions of the Royal Society of London 150, 57–84 (1860)
Maxwell, J.C.: XLIV. On physical lines of force Part II. Philosophical Magazine Series 4 21(140), 281–291 (1861)
Maxwell, J.C.: III. On physical lines of force Part III. Philosophical Magazine Series 4 23(151), 12–24 (1862)
Maxwell, J.C.: On Faraday's lines of force. Transactions of the Cambridge Philosophical Society 10, 27–83 (1864)
Maxwell, J.C.: Theory of Heat. Longmans, Green and Co., London (1872)
Maxwell, J.C.: A Treatise on Electricity and Magnetism, vol. II. Clarendon, Oxford (1873a)
Maxwell, J.C.: An essay on the mathematical principles of physics. By the Rev. James Challis, M.A., [Review]. Nature VIII, 279–280 (1873b)
Maxwell, J.C.: Matter and motion. Society for Promoting Christian Knowledge, London (1876a)
Maxwell, J.C.: On Bow's method of drawing diagrams in graphical statics, with illustrations from Peaucellier's linkage. Proceedings of the Cambridge Philosophical Society II, 407–414 (1876b)
Maxwell, J.C.: Constitution of Bodies. In: Baynes, S., Smith, W.R. (eds.) Encyclopædia Britannica, pp. 310–313. Scribner's Sons, New York (1878)
Maxwell, J.C.: Address to the Mathematical and Physical Sections of the British Association. In: Niven, W.D. (ed.) The Scientific Papers of James Clerk Maxwell, pp. 215–229. Cambridge University Press, Cambridge (1890)
Miller, D.W., Starr, M.K.: Executive decisions and operations research. Prentice-Hall, Englewood Cliffs (1969)
Peirce, C.S.: How to make our ideas clear. Popular Science Monthly 12, 286–302 (1878)
Rorty, R.: Objectivity, relativism, and truth. Cambridge University Press, Cambridge (1991)
Rubinstein, M.F.: Patterns of problem solving. Prentice-Hall, Englewood Cliffs (1974)
Sellars, W.: Science, Perception, Reality. Ridgeview Publishing Company, Atascadero (1963)
Shanks, D.: Hume on the Perception of Causality. Hume Studies 11(1), 94–108 (2011)
Shockley, W.: Transistor technology evokes new physics. In: Nobel Lectures, Physics 1942-1962, pp. 344–374. Nobel Foundation, Stockholm (1956/1998)
Thomson, J.J.: On the structure of the atom: an investigation of the stability and periods of oscillation of a number of corpuscles arranged at equal intervals around the circumference of a circle; with application of the results to the theory of atomic structure. Philosophical Magazine Series 6 7(39), 237–265 (1904)

Thomson, W.: Address by the President, Sir William Thomson, Knt., LL.D., F.R.S, pp. 84–105. John Murray, London (1872)

Thomson, W.: Inaugural address of the new President: Ether, electricity, and ponderable matter. Journal of the Institution of Electrical Engineers 18(77), 4–36 (1889)

Thomson, W.: The Size of Atoms: Popular Lectures and Addresses, p. 156. Macmillan and Co., London (1894a)

Thomson, W.: The Sorting Demon of Maxwell: Popular Lectures and Addresses, pp. 137–141. Macmillan and Co., London (1894b)

Thompson, S.P.: The Life of William Thomson, Baron Kelvin of Largs, vol. II. Macmillan and Co., London (1910)

Tyndall, J.: Address delivered before the British Association. Longmans, Green and Co., London (1874)

Tyndall, J.: Sound. A Course of Eight Lectures. Longmans, Green and Co., London (1867)

Visser, H.: Wittgenstein's Debt to Mach's Popular Scientific Lectures. Mind 91(361), 102–105 (1982)

Visser, H.: Boltzmann and Wittgenstein: Or How Pictures Became Linguistic. Synthese 119(1/2), 135–156 (1999)

Vorms, M.: Representing with imaginary models: Formats matter. Studies In History and Philosophy of Science Part A 42(2), 287–295 (2011)

Wittgenstein, L.: Remarks on the foundations of mathematics. von Wright, G.H., Rhees, R., Anscombe, G.E.M. (eds.) Anscombe, G.E.M (Trans.). Blackwell, Oxford (1967)

Wittgenstein, L.: Philosophical remarks. Rhees, R. (ed.) Hargreaves, R., White, R. (Trans.). Blackwell, Oxford (1976)

Wittgenstein, L.: Philosophical Investigations. Rhees, R (ed.) Anscombe, G.E.M. (Trans.). Blackwell, Oxford (1992)

Wittgenstein, L.: Culture and value: a selection from the posthumous remains. In: von Wright, G.H., Nyman, H., Pichler, A., Winch, P. (eds.) Winch, P. (Trans.). Blackwell, Oxford (1998)

Wu, H., Dougal, R., Jin, C.: Modeling power diode by combining the behavioral and the physical model. In: Franquelo, L.G., et al. (eds.) IECON 2005, pp. 685–690. IEEE Industrial Electronics Society, Piscataway (2005)

Chapter 2
Dimensional Analysis and Dimensional Reasoning

John Bissell

Imperial College of Science, Technology and Medicine, UK.

Abstract. This chapter explores some of the ways physical dimensions, such as length, mass and time, impact on the work of scientists and engineers. Two main themes are considered: dimensional analysis, which involves deriving algebraic expressions to relate quantities based on their dimensions; and dimensional reasoning, a more general and often more subtle approach to problem solving. The method of dimensional analysis is discussed both in terms of its practical application (including the derivation of physical formulae, the planning of experiments, and the investigation of self-similar systems and scale models) and its conceptual contribution. The connection between dimensions and the fundamental concept of orthogonality is also described. In addition to these important uses of dimensions, it is argued that dimensional reasoning (using dimensionless comparisons to simplify models, the application of dimensional homogeneity to check for algebraic consistency, and the 'mapping-out' of solutions in terms of parameter space) forms the implicit foundation of nearly all theoretical work and plays a central role in the way scientists and engineers think about problems and communicate ideas.

2.1 Introduction

> ...every particle of space is *always*, and every indivisible moment of duration is *everywhere*... (Isaac Newton 1687)

Measurement lies at the heart of both science and engineering. When we seek to make connections between our physical theories and experiment, or when we want to build a bridge or pumping station, we need to know the relationships between the quantities involved and - more specifically - their relative sizes. An experiment to test a theory of speed, for example, is not possible without some method of specifying relative distance, while a new railway bridge will be rightly considered a failure if it is not long enough to cover the relevant span. We are only able to practise science and engineering, therefore, by employing a system for

C. Bissell and C. Dillon (Eds.): Ways of Thinking, Ways of Seeing, ACES 1, pp. 29–45.
springerlink.com

communicating the relative magnitude of physical quantities. And it is this that necessitates measurement.

But what, exactly, does measurement involve? I could, if I so desired, use a ruler to measure the height in centimetres of the seat of my chair, finding this to be - say - 40cm. In the process I would be measuring a ratio: the ratio of the height of my chair to one centimetre, that is, 40:1. In doing so I have used a unit, in this case centimetres, as a base reference, but something more subtle has also taken place: I have made use of *dimensions*.

The dimensions of a measurement are independent of the units used as the basis of reference. Indeed, when measuring the height of my chair I could have used inches or feet, or even - should the distance have seemed especially large - light-years. Whatever the chosen unit, however, it would have to have been one appropriate for the dimension measured, in this case *length*. Exactly what length *is* constitutes something of an ontological problem, but for the time being I will state that it is the property of a system or object that confers upon that system or object extension in space. But length is not the only dimension of interest to a scientist or engineer. We must also make use of other dimensions; for example, *time*, which confers temporal extension, and *mass*. What should be noted here, as Newton observes in the opening quotation (Newton, 1687), is that each of these is in some sense fundamentally unique and cannot be expressed in terms of the others. And this difference, which allows the scientist or engineer to draw both distinctions between properties and to find relations between them, has profound consequences.

In this chapter I describe some of the important ways in which dimensions impact on, and are used in, the work of scientists and engineers. In doing so I will make a distinction between what I consider to be two key branches. First, there is *dimensional analysis*: explicit reasoning from the dimensional nature of elements in a set of physical quantities to algebraic relationships between those quantities; and second, what might be referred to as *dimensional reasoning*: the more implicit process of making comparisons between quantities of identical dimension, or of 'mapping out' a set of physical solutions in terms of dimensionless parameters. The first sense is possibly the one we are thinking of when we remember the *a priori* arguments taught at school or during undergraduate study. Nevertheless, the second, which makes up such an important part of the tacit skill base of those practising science and engineering, is perhaps that used most commonly.

What follows is broadly speaking divided along these lines. The latter half of our discussion will centre on a review of the method of dimensional analysis, consideration of its relationship to other forms of reasoning, and its application to both science and engineering. Though not intended as a primer in the method,[1] the inclusion here of a few worked examples will help to illustrate several discursive points. In principle, there is substantial philosophical work in this area. A thorough study of the issues involved - which range from 'fundamental dimensions', the place of dimensional argument in scientific explanation, to how and why such arguments work at all - is beyond the scope of this chapter, and

[1] Plenty of excellent books exist on this topic, not least *Dimensional Analysis* (Huntley 1953) and *Dimensional Analysis for Engineers* (Taylor 1974).

will only be briefly touched on. However, for the purpose of trying to uncover just what a scientist or engineer does, it is perhaps dimensional reasoning that is most important, and the first half of our discussion shall begin here. Many of the themes in this section may at first appear to 'state the obvious', but they are also key elements of the cultural and working knowledge that underpins scientific research.

2.2 Dimensional Reasoning

The dimensions length, mass and time, denoted by the letters L, M and T respectively, have already been introduced. In mechanical problems the set L, M and T form what we might call *fundamental dimensions*, so that the dimension of a given mechanical quantity must be expressed as some combination of these three. For example, when we consider the velocity of a particle v, we are interested in the number of length units the particle covers in a given number of time units. Thus the dimensions of velocity, written using the square bracket notation $[v] = V$, is length L divided by time T, that is, $V = LT^{-1}$. Because this expression gives the dimensions of V it is called the *dimensional formula* for V. Similarly, the dimensional formula for the particle's momentum $p = mv$, where m is its mass, is given by $[p] = MV = MLT^{-1}$, while the expression for its kinetic energy $E = \frac{1}{2}mv^2$ is $[E] = ML^2T^{-2}$.

Just which set of dimensions should be deemed generally fundamental is something of an open question. In the examples given above we could in principle take L, M and V as fundamental, yielding a dimensional formula for time $T = LV^{-1}$. That we don't probably has as much to do with the psychology of human beings, and our experiential understanding of the natural world from a 'common sense' perspective, as it does with the absolute nature of things. Indeed, it seems likely that scientists and engineers simply adopt whichever set seems most appropriate to the systems they study, just as they might choose an appropriate set of base units for measurement.[2]

Nevertheless, the choice of fundamental units has been a source of controversy, especially when that choice impacts on the solubility of a problem. In electrical systems, for instance, the necessity of introducing an additional dimension of charge Q is clear. However, in cases involving heat, defining a fundamental temperature dimension θ is more complicated. Indeed, there are some heat-flow problems that are only amenable to dimensional analysis provided θ is used in addition to L, M and T; yet from the perspective of kinetic theory, temperature is considered a form of kinetic energy with dimensions ML^2T^{-2}, rendering the same problem intractable (Rayleigh 1915a). As Lord Rayleigh notes, the situation is somewhat perplexing:

---

[2] For a discussion of the history of dimensional analysis and the connection between fundamental dimensions and base units, see *The Mathematics of Measurement: A Critical History* by J. J. Roche (Roche 1998).

> It would indeed be a paradox if the further knowledge of the nature of heat afforded by molecular [kinetic] theory put us in a worse position than before in dealing with a particular problem. (Rayleigh 1915b)

Unfortunately Rayleigh doesn't offer a resolution, though he speculates that there is perhaps something qualitatively different about the nature of thermal kinetic energy which warrants the use of θ in certain contexts. From a kinetic perspective this has some justification: the thermal energy associated with a collection of particles is proportional to the *mean of the square* of their random velocities, not the square of their mean velocity. However, this argument on its own seems rather unsatisfactory. As Huntley suggests, it may be that 'the criterion [for fundamentality] is purely pragmatic' (Huntley 1953). Nevertheless, before considering the role of fundamentality in the method of dimensions, I shall review some of the important ways in which *dimensional reasoning* in general shapes the work of scientists and engineers.

2.2.1 Dimensional Homogeneity and Commensurability

Dimensional reasoning involves considering the relationship between physical quantities based on the dimensional properties of those quantities alone. It relies on the principle of *dimensional homogeneity*: every term in a physical equation must have the same dimensional formula. When a physical quantity q is expressed as a sum of n other quantities p_i, such that $q = p_1 + p_2 + \ldots + p_n$, this means $[q] = [p_i]$. Similarly, if a quantity q is proportional to the product of n other quantities p_i themselves raised to powers a_i, then

$$q = Cp_1^{a_1} p_2^{a_2} \cdots p_n^{a_n} \Rightarrow [q] = [p_1]^{a_1} [p_2]^{a_2} \cdots [p_n]^{a_n}, \qquad (2.1)$$

where C is a dimensionless constant of proportionality. Dimensional homogeneity makes intuitive sense to the extent that, for example, the total momentum of a system is the sum of constituent momenta in the system, not the constituent velocities or the constituent masses. In this way, dimensional reasoning comes into play whenever we use physical formulae and is perhaps its most prevalent application. Certainly, dimensional homogeneity is an invaluable tool in the day to day work of both scientists and engineers, and is employed whenever examining the dimensional consistency of terms in an equation (as a quick test for algebraic errors) or trying to recall half-forgotten formulae.

Reasoning of this kind is also essential when comparing quantities: only those with the same dimensional formula are commensurable. Making magnitude comparisons between quantities whose dimensions differ is not meaningful; for instance, there is no sense in which 2 kilograms are greater than 1 metre (length and mass are said to be *incommensurable*). However, knowing which variables dominate physical processes is important to our understanding of how a system operates. Fortunately, quantities can be compared if they are expressed as dimensionless ratios. In a perturbation analysis, for example, we might wish to consider the relative importance of two quantities δx and δy perturbed from corresponding variables x_0 and y_0. If the x and y variables have different dimensions this is not possible; however, finding dimensionless orderings, such as $\delta x/x : \delta y/y$, can reveal which effect is most important.

Similarly, we can make comparisons between units of identical dimension. And by forming derived units, such as characteristic time and length scales, scientists and engineers can employ dimensional commensurability to think about the validity of their models. For example, in a system of non-uniform temperature T_e we can use the dimensions of the gradient operator ∇, which are $[\nabla] = \mathrm{L}^{-1}$, to define a characteristic length-scale $l_T = T_e / |\nabla T_e|$ over which the temperature varies by T_e. This length scale may then be compared to other relevant length scales in the system. Indeed, a key parameter in plasma physics is the electron thermal mean-free-path λ_T, the mean distance an electron travels before undergoing a collision. Broadly speaking, if $\lambda_T < l_T$, then we may assume that electrons deposit the thermal energy associated with their motion locally: in a region at approximately the same temperature as the region from which they originated (Braginskii 1965). In this case a *local* model of heat transport may be used. On the other hand, if $\lambda_T > l_T$, then electrons will stream rapidly into regions where the temperature differs considerably from that of their origin and a *non-local* model of heat transport is necessary. The value of the dimensionless parameter l_T/λ_T may then be used by plasma physicists to quickly characterise the nature of the heat-flow in the systems they study.

2.2.2 Model Simplification

A related use of dimensional comparisons concerns the simplification of mathematical expressions in theoretical work, which I shall illustrate using a contrived example, taken again from plasma physics. Consider the following expression for the rate of change of a vector function **f** in a fluid system with bulk flow velocity **C** and mean electron collision time τ_T :

$$\frac{\partial \mathbf{f}}{\partial t} = -\frac{\mathbf{f}}{\tau_T} + (\nabla \mathbf{C}) \cdot \mathbf{f}, \tag{2.2}$$

where $\nabla\mathbf{C}$ is a dyadic gradient operation yielding a rank-two tensor (matrix). Superficially, equations such as these may seem imposing due to the difficulty of interpreting the dyadic term; especially if the analytical form of **C** is either unknown or else fiendishly complicated. Nevertheless, we can make progress by combining physical understanding of the system and effective use of dimensional reasoning.

We proceed by noticing that for many systems the components of $\nabla\mathbf{C}$ are likely to be of a similar order of magnitude as $|\nabla C|$, with $C = |\mathbf{C}|$ as the plasma's bulk speed. This means that

$$\mathrm{O}\big((\nabla \mathbf{C}) \cdot \mathbf{f}\big) \sim \mathrm{O}\big(C(|\nabla C|/C)\mathbf{f}\big) \tag{2.3}$$

where the symbols 'O' and '~' mean *'of the order'* and *'similar to'* respectively.

Furthermore, we observe using dimensional reasoning (as we did with temperature T_e length scales above) that the characteristic speed length scale is given by $l_C = C/|\nabla C|$. The quantity $t_C = l_C/C$ thus represents the time taken for the bulk plasma to traverse a distance equal to that over which the magnitude of C changes by C. Hence, with reference to equation (2.3), equation (2.2) may be written

$$\frac{\partial \mathbf{f}}{\partial t} = -\frac{\mathbf{f}}{\tau_T} + \mathrm{O}\left(\frac{\mathbf{f}}{t_C}\right) \Rightarrow \frac{\partial \mathbf{f}}{\partial t} \approx -\frac{\mathbf{f}}{\tau_T} \tag{2.4}$$

where the implied approximation is justified for many plasmas on the basis that the collision time τ_T is generally much shorter than t_C: that is, $\mathrm{O}(\mathbf{f}/t_C)$ may be neglected when compared with $\mathbf{f}/\tau_T$.

Crucially, by using our physical intuition, and a small degree of dimensional reasoning about the relevant length and timescales, we have successfully simplified an otherwise fairly opaque equation with relative ease. Whether or not equation (2.4) itself is soluble is not at stake here, the point is that we have made progress towards a solution. Such approaches are invaluable in theoretical work: exact expressions have a tendency to be mathematically intractable; approximate forms are often far more amenable to analysis.

2.3 Method of Dimensional Analysis

Having addressed some of the more general uses of dimensional reasoning, we are now in a position to explore the rather elegant topic of dimensional analysis proper. An example is instructive here and we shall introduce the method by considering the classic problem of a simple pendulum. This is a useful example for two reasons: firstly, it illustrates well the essential features of the method; and secondly, it highlights some of the method's conceptual peculiarities.

2.3.1 Analysis of the Simple Pendulum

We begin by formulating an abstracted model for the pendulum, and to do this physical reasoning and experience must first be employed to emphasise certain features of the system compared to others. This is a process situated at the core of mathematical modelling, and the technique used when deriving equation (2.4) above. The system is constructed as follows: a bob of mass m is suspended from a light string of length l and displaced so that the string and vertical define a small angle ϕ (see Fig. 2.1). Due to the restorative force of the bob's weight mg, where g is the acceleration due to gravity, the pendulum will swing back and forth with a fixed time period t. It is an expression for t that we wish to obtain.

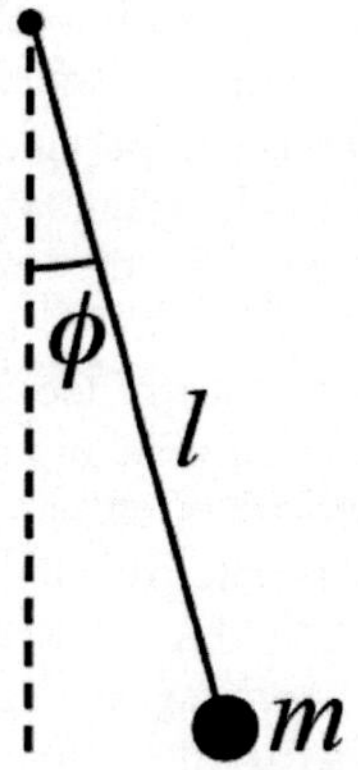

Fig. 2.1 The simple pendulum formed by suspending a bob of mass *m* on a string of length *l*. The angle ϕ between the string (solid line) and the vertical (dashed line) is assumed to be small.

The beauty of dimensional analysis is that this may be done without reference to either force balance or the solutions to differential equations. Indeed, the approach requires us simply to make an informed guess as to which quantities in the system are important and then proceed using basic arithmetic. Let us suppose that *t* depends only on the pendulum's length, which has dimensions L, the bob's mass, which has dimensions M, and the acceleration due to gravity, which has dimensions LT^{-2}.

Using what has been dubbed the *Rayleigh method*, we assume that *t* is proportional, by a dimensionless constant *C*, to the product of powers of *l*, *m* and *g*:

$$t = Cl^{\alpha} m^{\beta} g^{\gamma}, \tag{2.5}$$

where α, β and γ are constant exponents. Hence, by the expression for dimensional homogeneity of equation (2.1) we have:

$$[t] = [l]^{\alpha}[m]^{\beta}[g]^{\gamma} \Rightarrow \mathrm{T} = \mathrm{L}^{\alpha+\gamma}\mathrm{M}^{\beta}\mathrm{T}^{-2\gamma} \tag{2.6}$$

so that comparing the exponents on either side of equation (2.6) we find

exponents of T:	$1 = -2\gamma,$
exponents of L:	$0 = \alpha + \gamma,$
exponents of M:	$0 = \beta.$

Thus, solving for α and γ to find $\alpha = -\gamma = 1/2$, and substituting these values into equation (2.5) we have

$$t = C\left(\frac{l}{g}\right)^{1/2} \tag{2.7}$$

Notice that dimensional analysis does not tell us the value of C (in this case 2π) which we would have to obtain from experiment or a different type of mathematical analysis; neither does it describe the importance of the angle ϕ, though alternative approaches do provide some insight (Huntley 1952). However, the method has revealed some important features of the system, such as the time period's proportionality to the root of the pendulum's length, inverse proportionality to the root of g, and independence on the mass of the bob m.

Indeed, dimensional argument differs from other forms of physical reasoning in a number of ways. It does not offer causal descriptions of phenomena and is unable, without supplementary experimentation, to furnish us with accurately predictive equations. Its status, therefore, as a means of explanation, or tool for dealing with physical problems requires some exploration.

2.3.2 Conceptual Value

As a convenient route into a problem, especially when the underlying phenomenon is somehow obscured, dimensional analysis can prove unexpectedly powerful, and often enables scientists and engineers to bypass a more complicated mathematical treatment. Some of the classic applications of the method have been of this form: Rayleigh's initial exploration of why the sky is blue, for example, does not recourse to full-blown electrodynamic modelling of interactions between incident light from the sun and scattering particles in the upper atmosphere (Rayleigh 1871a; 1871b). Indeed, in the space of two short paragraphs he argues that the ratio of the amplitudes of scattered and incident light R_A is proportional to a dimensionless function of three quantities: its wavelength λ, which has dimensions L; the distance to the scattering particle r, which also has dimensions L; and the particle's volume V, which has dimensions L^3. The relationship $R_A \propto V/r$ was already known to Rayleigh, so that r, λ, and V may only be combined in a dimensionless fashion to give $R_A \propto (V/r\lambda^2)$ and the key result:

> ...the ratio of the amplitudes of the vibrations of the scattered and incident light varies inversely as the square of the wave-length, and the intensity of the lights themselves as the inverse fourth power. (Rayleigh 1871a)

Thus, since blue light has a wavelength approximately half that of red light, it is more strongly scattered than the latter by a factor of ~16: it is this blue scattered light that we see when looking at the sky.[3]

Indeed, though not able to provide complete solutions, the fact that dimensional analysis can supply key scalings is invaluable. In the example of the pendulum, the unknown constant of proportionality between t and $(l/g)^{1/2}$ is in many ways of less interest than the functional dependence of t on l and g. To the physicist or engineer a working knowledge of such dependence is often of more use than a numerically accurate expression. If, for example, we want to double the period of a given pendulum, equation (2.7) tells us that we must quadruple its length; this

[3] That we see blue light rather than purple is a result of additional effects, including the physiology of the human eye.

reveals more about pendulums generally than, after having l, g and the 2π proportionality supplied, calculating the value of t for a specific case.

Knowledge of relevant scaling is particularly important when trying to conserve experimental effort, as the following example adapted from Huntley demonstrates (Huntley 1952). We suppose that an experiment is designed to study the dependence of a quantity q on other variables r, s and t. During the experimental planning stage, a brief dimensional analysis yields

$$q = \frac{r}{t^2}\left(\frac{s}{t}\right)^{\alpha} \tag{2.8}$$

where α is an unknown exponent to be determined.

Our *a priori* expression for q tells us that an experiment to measure the change of q with r, though enabling us to determine the constant of proportionality, would not be the most fruitful choice of investigation. It would be much better, in this case, to consider the change of q with s or t, and in doing so determine the value of α. We might, for example discover that $\alpha = -2$, implying that q is independent of t and proportional to r/s^2. When detailed theoretical work is either intractable or obfuscatory, the method of dimensions can support the work of scientists and engineers, acting as a guide in experimental endeavour.

Nevertheless, dimensional analysis has an additional, and more subtle conceptual value prompting different ways of responding to a physical solution. In the pendulum example we began by explicitly selecting what we thought were the principal variables of the system, namely l, g and m, only to go on to exclude m from appearing in our expression for t. In this way, the analysis tells us very clearly that the time period of a pendulum *cannot* depend on m. Alternatively, had we begun by arguing from Newton's laws of motion, the cancellation of m in our derivation may well have passed unnoticed as an instance of mathematical felicity. In the first case we set out what we believed to be important, discovering both what *is* and what *is not* by comparison with our initial supposition. In the second case no such comparison is available because no initial supposition is made. Conceptually we learn different things about the system depending on our method of analysis, though these methods are, of course, complementary. By following the dimensional method we discover that there is no functional dependence of t on m; turning to a forces-based argument we can find out why.

2.3.3 Dimensions and Orthogonality

Before proceeding further it is worth considering in greater detail how dimensional analysis and dimensional reasoning work, and what their working means for our understanding of physical problems. As we have seen, the basic principle of the analysis is dimensional homogeneity, but that this principle should be effective at all is a consequence of its relationship to one of the key concepts in science and engineering: orthogonality.

There are a number of ways of thinking about orthogonality. However, the one most appropriate to dimensional analysis and dimensional reasoning is the orthogonality of vectors (though it is not clear that dimensions themselves form a vector space in the proper sense). For example, displacement is a scalar quantity, with an absolute magnitude reflecting the distance displaced; while the change in position following a displacement is a vector quantity, reflecting the magnitude of displacement and direction in which the displacement occurs. In an orthogonal system the different components of a vector act independently. For example, following a displacement in a Cartesian (x, y) geometry, the change in the x-position tells us nothing about the change in y; information about how both vary is needed for a complete description.

In a similar fashion, the dimensions of a quantity vary independently. When we compared exponents of L, T and M in equation (2.6), we were able to do so because any exponent change in a given dimension on one side of the equation had to be accounted for by an equal change in the same dimension on the other. Indeed, this is the basis of dimensional homogeneity discussed earlier, which in some respects represents one of the most fundamental manifestations of orthogonality in natural science.

The connection between orthogonality and dimensions may be made more explicit by introducing the concept of *vector length*.[4] Treating length as a scalar we may denote its singular dimension as L; while treating it as a vector we find it has three component dimensions L_x, L_y and L_z, where the subscript carries information about the (Cartesian) direction considered. Furthermore, these definitions allow us to treat the dimensions of other quantities with greater descriptive precision. In a vector system, for example, velocity in the x-direction may be more accurately given the dimensions of L_xT^{-1}. And as we shall see in the following section, this approach greatly enhances the power of dimensional analysis.

2.4 Scale Models, Similarity and Scaling Laws

The method of dimensions can be particularly important to engineers during the process of designing prototypes, and has long been used as a technique for thinking about the relationship between scale models and what those models actually represent. Here it is the *principle of similitude* which is important; the idea that the responses of different sized models may be compared because relevant physical phenomena act in the same way over a range of scales. In the past, these approaches have been widely applied to problems involving the motion of bodies in fluids; systems for which the number of variables, and the complexity of the equations relating them, makes general solution impossible. Indeed, such cases are often of considerable technological importance: when designing an aircraft, for example, it is necessary to know the impact of air resistance when planning the number and power of its engines, and the size of its fuel tank.

[4] Huntley provides a comprehensive development of this approach (Huntley 1952), attributing its origin to much earlier work in the late nineteenth century (Williams 1892).

Aircraft design is one of the classic applications of dimensional analysis, and serves as a useful example of how scaling laws derived from the method may be applied in a practical setting. To illustrate this we shall consider a slightly artificial scenario, again adapted from Huntley (Huntley 1952), making use of the vector length dimensions previously mentioned. Symmetry considerations, however, mean that Huntley's analysis can be simplified from the set of three vector lengths (L_x, L_y and L_z) to two.

Suppose that we wish to predict the resistance R likely to be met by a prototype aircraft using measurements derived from a scale model in a wind tunnel. Our first task is to find an expression for the air resistance to a model placed in a medium (air) of density ρ, viscosity η and flowing with velocity v. Since, the direction of flow defines only one unique axis (about which solutions are invariant under rotation) we require only two fundamental length dimensions: those parallel to flow $L_{\parallel}$ and those perpendicular $L_{\perp}$. In this way, v has dimensions $L_{\parallel}T^{-1}$, while the model's characteristic length l and cross-section A have dimensions $L_{\parallel}$ and $L_{\perp}^2$ respectively.[5] The quantities and their dimensions are tabulated below:

Table 2.1 Physical quantities and their dimensions.

Quantity q	**Dimensions** $[q]$							
Resistance R	M	×	$L_{\parallel}$	-	-	×	T^{-2}
Flow velocity v	-	-	$L_{\parallel}$	-	-	×	T^{-1}
Density of medium ρ	M	×	$L_{\parallel}^{-1}$	×	$L_{\perp}^{-2}$	-	-
Characteristic length l	-	-	$L_{\parallel}$	-		-	-
Characteristic cross-section A	-	-	-	-	$L_{\perp}^{2}$	-	-
Viscosity of medium η	M	×	$L_{\parallel}^{-1}$	-	-	×	T^{-1}

As before, we assume that R may be expressed as the product of the other quantities raised to fixed powers α, β, γ, χ and δ, multiplied by a dimensionless constant of proportionality C. In this way we have

$$R = Cv^{\alpha}\rho^{\beta}l^{\gamma}A^{\chi}\eta^{\delta} \tag{2.9}$$

Performing the analysis we find that R cannot be unequivocally determined, so that

$$R = Cv^2\rho A\left(\frac{\rho v A}{l\eta}\right)^{-\delta} \tag{2.10}$$

where δ is unknown.

The dimensionless quantity $\rho vA/l\eta$ is in fact Reynold's number, a key parameter in fluid mechanics. That δ should remain undetermined reflects the fact that scaling with $\rho vA/l\eta$ is dependent on the peculiarities of a given problem, a

[5] If flow is in the x-direction, for example, this may be thought of as setting $L_x = L_{\parallel}$ and $L_{\perp} = (L_yL_z)^{1/2}$, so that $[l] = L_{\parallel} = L_x$ and $[A] = L_{\perp}^2 = L_yL_z$.

point we shall return to. However, for current purposes we note that comparison between the model and the prototype is possible without knowing δ provided we ensure $\rho vA/l\eta$ takes the same value in both systems. The process of enforcing identical values for dimensionless quantities in equations for systems on different scales is referred to as *dimensional similarity*, and may be understood as follows.

Denoting quantities associated with the model using the subscript m and those associated with the prototype using the subscript p, and assuming that the model is on a scale 1:r, the characteristic length scales and cross-sections are such that $l_p/l_m = r$ and $A_p/A_m = r^2$. Then, since the density and viscosity of air is the same for both model and prototype, that is, $\eta_m = \eta_p = \eta$ and $\rho_m = \rho_p = \rho$, we find

$$\frac{R_p}{R_m} = \left(\frac{v_p}{v_m}\right)^2 r^2 \left(\frac{\rho v_p A_p}{l_p \eta}\right)^{-\delta} \left(\frac{l_m \eta}{\rho v_m A_m}\right)^{-\delta}. \tag{2.11}$$

Notice that δ remains problematic because the dimensionless quantity $\rho vA/l\eta$ has not yet been fixed to take the same value on both scales. For this condition to hold, the model must be tested under conditions such that $v_m = rv_p$, in which case

$$\frac{\rho v_m A_m}{l_m \eta} = \frac{\rho(rv_p)(A_p/r^2)}{(l_p/r)\eta} = \frac{\rho v_p A_p}{l_p \eta} \tag{2.12}$$

and equation (2.11) reduces simply to the fraction

$$\frac{R_p}{R_m} = 1. \tag{2.13}$$

Hence, by measuring R_m we indirectly arrive at R_p. Indeed, what we have discovered is a *scaling law* between the model and the prototype: the resistance encountered by the prototype travelling with velocity v_p, is identical to that of the model in a wind tunnel with wind-speed $v_m = rv_p$. Explicitly stating the velocity argument in our notation for resistance, so that we use $R_p(v_p)$ for resistance in the former case and $R_m(v_m)$ for the latter, this scaling law may be expressed more concisely as

$$R_p(v_p) = R_m(v_m = rv_p). \tag{2.14}$$

Our example of the model aircraft is instructive because it clearly demonstrates the power of dimensional analysis when applied to problems of engineering in complicated systems such as fluids. It should be noted, however, that there is some artificiality to the problem in this instance. Given that aircraft must often travel at very high speeds, it may well be impossible to reach velocities of rv_p for the model. Even for large models, with $r \approx 10$ for example, wind-tunnel speeds of thousands of miles per hour would be necessary.

Nevertheless, aircraft designers can again rely on dimensional analysis for assistance. Supposing we found by experimentation with our model that $\rho vA/l\eta$ tends asymptotically towards a fixed value as v is increased. For sufficiently high speeds, therefore, we find that resistance becomes proportional to the square of the velocity, that is, equation (2.10) may be approximated as

$$R(v) \approx C' v^2 \rho A, \tag{2.15}$$

where C' is the product of C with the asymptotic value of $\rho vA/l\eta$. Hence, combining this approximate expression with equation (2.14), it may be shown that

$$R_p(v_p) = \frac{R_m(rv_p)}{R_m(v_m)} R_m(v_m) \approx \left(\frac{v_p}{v_m}\right)^2 r^2 R_m(v_m). \tag{2.16}$$

In this way, investigation of the functional dependence of $\rho v_m A/l\eta$ with v_m provides us with a means of finding R_p even when wind tunnel speeds of rv_p are unobtainable.

However, the use of scaling laws and dimensional similarity is not restricted to design problems involving scale models. Indeed, in the 1940s Sir Geoffrey Taylor famously used dimensional scaling to discover the otherwise undisclosed yield of the United States' then most advanced atomic weapon (Taylor 1950a; 1950b).[6] Following its detonation at time t, in air of atmospheric density ρ, the energy E of a bomb may be related to the radius of its blast wave r using the Rayleigh method:

$$\begin{aligned} &[E] = [r]^{\alpha}[t]^{\beta}[\rho]^{\gamma} \Rightarrow \mathrm{ML^2T^{-2}} = \mathrm{M^{\gamma}L^{\alpha-3\gamma}T^{\beta}} \Rightarrow E = C\frac{\rho r^5}{t^2} \\ &\Rightarrow \log r = \frac{2}{5}\log t + \frac{1}{5}\log E - \frac{1}{5}\log\rho - \frac{1}{5}\log C \end{aligned} \tag{2.17}$$

Here I have disguised some of the complexity of Taylor's analysis (which involved solving a self-similar shock problem, not the simpler Rayleigh method) in the unknown dimensionless quantity C.[7] The key point is that his solutions were dimensionally similar and he was able to determine $C \approx 1$ using scaling laws from smaller explosion trials of known energy. Labelled photographic snapshots of the atomic test were not secret (they had been declassified in 1947 and clearly indicated the radius of the blast in successive time frames), so it was then straightforward for Taylor to determine E from the intercept of the logarithmic plot of r against t. Taylor's estimate of the yield at 16,800 Tons is impressively close to the released figure of 20,000 Tons.

In more recent years, scaling laws have been successfully applied to experiment with laboratory astrophysics (see, for example, Falize et al. (2009)). Since actual astrophysical systems are well beyond the possibility of experimental replication on a 1:1 scale, dimensional similarity is a key method of investigation for scientists seeking repeatable tests of their theories.

---

[6] These papers were not, so the story goes, popular with the U.S. military.

[7] I have also clearly used dimensional quantities inside the logarithms and the reader must assume that these are normalised by the units in which they are measured.

2.4.1 The Π-Theorem

No discussion of the use of dimensional analysis would be complete without mentioning what has become known as the Π-Theorem - the seminal formal work on the topic published by E. Buckingham in the early 20th Century (Buckingham 1914). Broadly speaking, the theorem states that a dimensionally homogeneous equation involving n variables, defined in terms of k orthogonal dimensions, may be reduced to an equation in $(n - k)$ dimensionless parameters Π_1, Π_2, ..., Π_{n-k}. Together, these parameters characterise the system of interest and a relationship between them obtains such that:

$$\Pi_1 = \varphi_1(\Pi_2, \Pi_3, \cdots, \Pi_{n-k}), \tag{2.18}$$

where the function φ_1 is unknown and the subscripts are interchangeable. From this expression, further relations between the dimensionless quantities and their constituent variables may be derived in a similar fashion to the manner in which we proceeded earlier. For example, in the problem of the model aircraft we have six variables (R, v, ρ, l, η, A) and four orthogonal dimensions (M, $L_{\parallel}$, $L_{\perp}$, T), so that n = 6 and k = 4. The system may thus be characterised by the $n - k = 2$ parameters $\Pi_1 = (R/v^2\rho A)$ and $\Pi_2 = (\rho vA/l\eta)$. In this way equation (2.18) allows us to write

$$\frac{R}{v^2\rho A} = \varphi(\rho vA/l\eta), \tag{2.19}$$

which the reader may recognise as an alternative form of equation (2.10). Naturally there are subtleties to the Π-Theorem which have been glossed over here, but our result shows something important about the scope of dimensional analysis by making it clear that the functional dependence is not necessarily with exponent. Indeed, the undetermined value of δ in equation (2.10) was a consequence of this fact: δ is not simply a constant and the relationship of R to $(\rho vA/l\eta)$ is more complicated than a power law.

Nevertheless, as we saw with the scale model, the details of the function φ are not necessarily crucial so long as it may be approximated for a range of conditions. Indeed, the undetermined aspect of the analysis is often of benefit to scientists and engineers because it can reveal additional features of the underlying system. For the case considered, the dependence on $\rho vA/l\eta$ in a given regime will, more-or-less, follow a single power law; a shift towards a different power, therefore, reflects changes to the principal dynamics and transition to a different regime. This quality was born out in our treatment of model aircraft: the asymptotic approach of $\rho vA/l\eta$ towards a constant corresponds to the rapidly decreasing impact of viscosity, when compared to turbulence, in the high velocity regime (Huntley 1952).

2.4.2 Dimensionless Parameters and Solution Space

The discussion of dimensionless parameters in the last topic returns us neatly to the more general consideration of dimensional reasoning in science and engineering that we began with. As we have seen, scientists and engineers typically use dimensionless parameters to 'map out' different regimes in the systems they study. These techniques become particularly powerful when equations are re-written in terms of dimensionless parameters and expressed graphically.

As an example, consider a plasma instability (Bissell 2010) whose maximum growth rate γ_M may be approximated by

$$\gamma_M = \varphi(\chi)\frac{1}{\tau_T}\left(\frac{\lambda_T}{l_T}\right)^2, \tag{2.20}$$

where λ_T and l_T are the electron thermal mean-free-path and temperature length scale in the plasma mentioned earlier, τ_T is the mean time between collisions and $\varphi(\chi)$ may be taken to be constant.[8] Suppose we wanted to graphically represent the growth-rate as a function of the temperature length scale. Naïvely, one might go about this by choosing values for λ_T and τ_T and plotting γ_M against l_T, thus showing the inverse proportionality inherent in equation (2.20). However, such a graph would only be immediately relevant to systems for which λ_T and τ_T took the same values as those used to construct the plot. It would be more interesting to use a dimensionless approach and plot the combined parameter ($\gamma_M \tau_T$) against (l_T/λ_T); this continues to reveal the inverse proportionality of γ_M and l_T, but has the added advantage of being applicable to systems for which a vast range of values λ_T and τ_T are relevant. It should be noted that a local model was used to derive equation (2.20),[9] so that the l_T/λ_T axis has the additional advantage of naturally identifying areas where the expression for the growth rate begins to break down, namely, where $l_T/\lambda_T < 1$ for which $\lambda_T > l_T$.

The use of dimensionless parameters as the scales on axes in graphics is often referred to as mapping out the graphic in *parameter space*, and is a convenient means of displaying more information than may be relayed on other types of plot. It is a standard technique in the day to day work of scientists and engineers who seek to quickly spot regimes where given phenomena are important, or rapidly communicate their ideas to others during discussion.

2.5 Conclusions

Throughout this chapter we have seen a number of ways scientists and engineers use dimensional analysis in their work and research: be that through the *a priori* derivation of physical formulae; as an initial aid when planning fruitful

[8] These quantities were discussed in Sect. 2.2.1 *Dimensional homogeneity and commensurability* in the paragraph on derived units.

[9] See previous footnote.

experimentation; or the use of scale models, dimensional similarity and scaling laws. The power of the method rests not only in its relative ease of implementation, but also in its versatility; with applications ranging from problems involving heat-flow (Rayleigh 1915) to the study of a cat's lapping mechanism when drinking (Reis *et al.* 2010). Furthermore, the analysis can also impact on the way scientists and engineers interpret the systems they study, by identifying key combinations of variables as dimensionless parameters which characterise behaviour in different regimes.

However, alongside these varied and important applications it is worth remembering that dimensional analysis itself has its origins in the more humble, but no less significant, field of dimensional reasoning. Indeed, arguably the most frequent use of dimensions by scientists and engineers is reasoning of this form. When trying to remember half-forgotten formulae, or when attempting to derive new ones, the scientist or engineer will frequently appeal to the principle of dimensional homogeneity as an aid to memory or technique for identifying algebraic errors. In fact, every time an equation is written down, quantities compared and models simplified, the principle of dimensional homogeneity is tacitly employed. Furthermore, through the use of parameter space and dimensionless ratios, dimensional reasoning plays a key role in generalising concepts beyond specific circumstances enhancing the interpretation, discussion and communication of ideas.

References

Bissell, J.J., et al.: Field Compressing Magnetothermal Instability in Laser Plasmas. Physical Review Letters 105(17), 175001 (2010)

Braginskii, S.I.: Transport processes in a plasma. Reviews of Plasma Physics 1, 205 (1965)

Buckingham, E.: On physically similar systems; illustrations of the use of dimensional equations. Physical Review 4(4), 345–376 (1914)

Falize, E., Bouquet, S., Michaut, C.: Scaling laws for radiating fluids: the pillar of laboratory astrophysics. Astrophysics and Space Science 322(1-4), 107–111 (2009)

Huntley, H.E.: Dimensional Analysis. MacDonald & Co. Ltd., London (1952)

Newton Sir, I.: The General Scholium appended to Philosophiae Naturales Principia Mathematica. Royal Society, London (1687); English edition, Sir Isaac Newton's Mathematical Principles of Natural Philosophy and His System of the World, translated by Andrew Motte, edited by R. T. Crawford, and revised by Florian Cajori (University of California, Berkeley (1934) 5th edn. (1962))

Rayleigh (Strutt, J. W., Third Baron Rayleigh): On the light from the sky, its polarisation and colour. Philosophical Magazine Series 4 41(271), 107–120 (1871a)

Rayleigh (Strutt, J.W., Third Baron Rayleigh): On the scattering of light by small particles. Philosophical Magazine Series 4 41(275), 447–454 (1871b)

Rayleigh (Strutt, J.W., Third Baron Rayleigh): The principle of similitude. Nature 95, 66–68 (1915a)

Rayleigh (Strutt, J.W., Third Baron Rayleigh): Letters to the Editor: The principle of similitude. Nature 95, 644 (1915b)

Roche, J.J.: The Mathematics of Measurement: A Critical History. The Athlone Press (Springer), London (1998)

Reis, P.M., Jung, S., Aristoff, J.M., Stocker, R.: How Cats Lap: Water Uptake by Felis catus. Science 330(6008), 1231–1234 (2010)

Taylor, E.S.: Dimensional Analysis for Engineers. Clarendon Press, Oxford (1974)

Taylor, G.: The Formation of a Blast Wave by a Very Intense Explosion I. Theoretical Discussion. Proceedings of the Royal Society A 201(1065), 159–174 (1950a)

Taylor, G.: The Formation of a Blast Wave by a Very Intense Explosion II. The Atomic Explosion of 1945. Proceedings of the Royal Society A 201(1065), 175–186 (1950b)

Williams, W.: On the relation of the dimensions of physical quantities to directions in space. Proceedings of the Physical Society 11, 357 (1890)

Chapter 3
Models: What Do Engineers See in Them?

Chris Dillon

The Open University, UK

Abstract. Developments in areas such as communications, control and switching systems, notably those moving into the public domain after the end of World War II, increasingly used complex mathematical ideas to model the way the systems behaved. However, not only was the mathematics unfamiliar to many engineers but, perhaps more significantly, the mathematics by itself often gave little or no insight into why a system behaved as it did, or how it might be designed to behave in a particular way. Rather than working directly with the mathematics, engineers developed ways of describing systems by constructing models that represented behaviour pictorially and which gave the experienced user a more immediate 'feel' for what was going on. It is proposed that this process is a key feature of the way engineers talk and think about the systems they design and build. Two examples from the evolving discourses in the communication and control engineering communities following World War II offer a view of this process in action. This chapter argues that the models engineers develop and use open up new ways of talking about systems that become part of the everyday language of communities of engineering practice.

3.1 Introduction

Developments in areas such as communications, control and switching systems, notably those moving into the public domain after the end of World War II, increasingly used complex mathematical ideas to model the way the systems behaved. However, not only was the mathematics unfamiliar to many engineers but, perhaps more significantly, the mathematics by itself often gave little or no insight into why a system behaved as it did, or how it might be designed to behave in a particular way. Rather than working directly with the mathematics, therefore, engineers developed ways of describing systems by constructing models that represented behaviour pictorially and which gave the experienced user a more immediate 'feel' for what was going on.

This chapter explores this idea in the context of two design techniques emerging after WWII. The first, relatively simple, technique was used to represent the

C. Bissell and C. Dillon (Eds.): Ways of Thinking, Ways of Seeing, ACES 1, pp. 47–69.
springerlink.com

logical switching functions required in telephone networks, sequence control of industrial processes and the embryonic digital computer industry. Following the work of Claude Shannon (1938) switching and logic circuits were characterised by expressions using Boolean algebra. However, Boolean algebra can be difficult to use and offers little insight into the design of efficient circuits. An approach introduced by Veitch (1952) and adapted by Karnaugh (1953), showed that the mathematics of Boolean algebra could be simplified by a diagrammatic representation, or map, where the mathematical manipulations of Boolean algebra were replaced by a simple pattern recognition process.

The second example is more complex and considers some of the graphical techniques used extensively to replace the mathematics of differential equations and complex algebra in the design of control systems from precision servomechanisms to large scale process plant. Based on the 1930s work of Nyquist, Black and Bode at the Bell Telephone Laboratories in developing frequency response methods for the design of feedback amplifiers, these techniques became standard tools for the engineers designing control systems in the 1940s and 50s. The story here is more than a simple linear narrative, however, and reveals a little of the flux of ideas as engineers on both sides of the Atlantic grappled with new concepts, terminology and techniques.

In this chapter I will argue that the models engineers developed and used were not just convenient representations of behaviour, rather the pictorial representations provided new ways of talking and thinking that became part of the language of the engineering community. I also want to suggest that that this process does not happen all at once but takes time to propagate and become accepted and shared practice.

3.1.1 Communities of Engineering Practice

The idea of a community of practice, introduced by Etienne Wenger and Jean Lave in the context of social theories of learning and subsequently further developed by Wenger (1998), has gained currency in the study of how groups of practitioners learn and create meaning for themselves. For Wenger the key features of a community of practice are captured by a *domain* of expertise, a *community* of practitioners that share and learn from each other, and a *practice* that has 'experiences, stories and tools' in common. This notion of a community that shares practice and evolves ways of talking about that practice fits engineering well. However, relatively little attention seems to have been paid to the ways engineers developed and represented ideas in their communities. Engineering practices often appear to be subsumed into more general 'technology' and then bundled into treatments of 'science and technology' where the emphasis is on scientific method and models that attempt to come to grips with the behaviour of the physical world for the purposes of description and analysis.

In contrast to science, engineering practice is explicitly focused on design – the process of constructing devices (which can be anything from simple components to complex systems and plant) that behave in specified ways. Edwin Layton, discussing the ideologies of science and engineering, observed:

> From the point of view of modern science, design is nothing, but from the point of view of engineering, design is everything. It represents the purposive adaptation of means to reach a preconceived end, the very essence of engineering. (Layton 1976, 696)

But this 'purposive adaptation of means' is not simply the mechanical application of rules. For Ferguson (1992) the notion of design involves a large component of visualization, and he argues that engineers carry with them information encoded in one form or another as images which they see in their 'mind's eye':

> Visual thinking is necessary in engineering. A major portion of engineering information is recorded and transmitted in a visual language that is in effect the lingua franca of engineers in the modern world. It is the language that permits 'readers' of technologically explicit and detailed drawings to visualise the forms, the proportions, and the interrelationships of the elements that make up the object depicted. It is the language in which designers explain to makers what they want them to construct. (Ferguson 1992, 41)

Ferguson is here talking largely of engineering drawings as a means of communicating between the design engineer and the skilled workers who will use the drawings to make real artefacts. However, the idea extends beyond the traditional relationship between designer and builder. Ferguson's 'visual language' need not be just mental pictures of the physical artefact when it is built, enabling 'designers [to] explain to makers what they want them to construct', but rather it becomes a powerful way of talking and communicating between engineers as they discuss and explain designs, techniques and resulting behaviours. Quoting a letter sent to him by the sociologist Kathryn Henderson concerning a study of the politics of engineering, Ferguson remarks:

> ...[she] remarked on the way talking sketches were made; she 'observed designers actually taking the pencil from one another as they talked and drew together on the same sketches'. Those talking sketches, spontaneously drawn during discussions with colleagues, will continue to be important in the process of going from vision to artefact. Such sketches make it easier to explain a technical point, because all parties in a discussion share a common graphical setting for the idea being debated. (Ferguson 1992, 97)

This learning and sharing of visual representations and what can be said about them shapes the way a community of engineering practice sees, thinks and talks. Going back to Wenger's notion of a community of practice he suggests that in any such established community we would expect to find three main features: an area of expertise, or at least common interests and competence; a community which offers support and help, where interests can be shared and discussed and where learning takes place; and a shared practice in which 'a set of stories and cases' support how the members function and interact on a day-to-day basis[1]. For engineers this shared practice includes a variety of ways to understand, visualise and communicate the behaviour of the systems they are designing and building, including charts, diagrams, sketches, graphs, engineering drawings, mathematical models and scale models. Different representations are used to give insight into how a system works; to explain behaviour and performance; to design and predict

[1] See http://www.ewenger.com/theory. Accessed 6 June 2011.

the behaviour of systems yet to be built; in short, to help tell a good story about what is going on.

Experienced engineers become adept at telling and sharing convincing stories within their communities. For Ferguson's and Henderson's engineering designers, then, we see much of this shared understanding in action as they talk about, sketch and modify their 'mind's eye' visions; passing ideas backwards and forwards in a mixture of technical jargon, quick sketches and back-of-envelope calculations. But this account appears to identify a fairly mature stage in the process of community formation. It describes a situation where the 'common graphical setting' has already been established, shared and understood to become part of 'normal' engineering. But where does all this understanding and commonality come from? How does the shared language evolve?

Engineers build up shared understandings of the system they are working with by embedding their models deeply in their conversations so that the boundaries between models and real world systems become blurred. In this way of talking engineering language evolves both from theory and from experience, so that features of the models become features of the real systems they represent. Morgan and Morrison (1999), focusing on scientific and economic models, suggest that models mediate between theory and the world as 'autonomous agents [and] instruments of investigation' (p. 10) but also that they 'are not passive instruments, they must be put to work, used, or manipulated' (p. 32). I will argue that models of engineered systems' behaviour play a similarly active role in mediating understanding, learning and conversation within the engineering community.

3.1.2 *Ways of Seeing*

To explore how ways of thinking and talking are linked to visual representations in engineering I am going to discuss two examples of engineering design techniques as they emerged in the late 1940s and early 1950s within the particular communities of control and communications engineers.

The first is a simple technique used to represent the relationship between input and output states in an electrical switching circuit. In this application engineers use a 'map' – a model of the logical relationships in a circuit that allows the engineer to identify and manipulate patterns of behaviour, rather than mathematical expressions.

The second is a more complex story which traces strands of the evolution of a graphical modelling technique from its origins in communications engineering in the 1930s to its use in the design of post-war control systems. The technique, called frequency response analysis, became a standard engineering tool during World War II in the design of servomechanisms for applications such as gun aiming, but following the war was also applied to the control of large scale processes, for example, in the petrochemical and power generation industries. However, new models, tools and techniques take time to become assimilated and embedded sufficiently deeply in a community to ensure common conventions and ways of talking. The coming together of ideas and practices from Europe and the United States created an environment in which informal learning between practitioners fuelled new ways of talking about systems. The evolving discourses in the engineering communities concerned with communication and control systems before and after World War II offer historians of technology a view of this process in action.

3.2 From Mathematics to Maps

Switching circuits were used extensively in telephone networks, control of industrial processes and, following WWII, in the growing digital computer industry. In 1938 Claude Shannon, later to become one of the pioneers of information theory (Shannon 1948), published a paper on the symbolic analysis of relay and switching circuits drawn from his MIT Masters thesis (Shannon 1938). He showed that Boolean algebra, introduced by George Boole in the study of propositional logic in the 19th century, could be used to describe relationships between inputs and outputs in systems where variables could take only two values. Originally used by Boole to represent logical propositions where the variables were either 'true' or 'false', Shannon applied the same approach to circuits made up of electromagnetic relays or switches, where a switch or a relay contact could be only 'open' or 'closed', and a voltage or current could be only 'on' or 'off'.

Applying the rules of Boolean algebra to circuits made up of interconnected switches led Shannon to express the 'state' of the output of a circuit, that is whether it was 'on' or 'off', as a Boolean equation. From an engineering perspective, however, what was powerful about this technique was that it allowed engineers not only to describe unambiguously what their circuits did, but also to specify uniquely the logical functions needed to perform a particular switching task and then design and build the circuit that performed the required function; for example, to ensure that interlocks on a machine prevented it from being operated unless safety guards were in place and a set start-up procedure was followed. Shannon described the design problem like this:

> In general, there is a certain set of independent variables A, B, C ... which may be switches, externally operated or protective relays. There is also a set of dependent variables x, y, z ... which represent relays, motors or other devices to be controlled by the circuit. It is required to find a network which gives, for each possible combination of values of the independent variables, the correct values for all the dependent variables. (Shannon 1938, 721)

Shannon's method enabled engineers to build mathematical models of their switching systems. These models were in the form of Boolean equations which expressed the outputs of a circuit (the dependent variables) in terms of the logical states of the inputs (the independent variables). A typical Boolean equation might look like:

$$X = ABC + \overline{A}BC + A\overline{B}\overline{C} + AB\overline{C}$$

A, *B* and *C* represent inputs, such as switch contacts, that have a 'truth' value of 1 or 0, depending on whether they are open or closed. So, for example, if switch A is closed then $A = 1$; if switch B is open then $B = 0$. The elements $\bar{A}$, $\bar{B}$ and $\bar{C}$ are the logical inverses of the inputs such that if $A = 1$ then $\bar{A} = 0$, or if $A = 0$ then $\bar{A} = 1$; and similarly for B and C. The products ABC, $\bar{A}BC$, $A\bar{B}\bar{C}$ and $AB\bar{C}$ represent logical AND functions of the inputs and their inverses, and the '+' signs separating the products represent logical OR functions. The output X is either 1 or 0, on or off, depending on the combination of the input states[2]. As with conventional algebra Boolean algebra allows expressions like this to be manipulated so that terms can be rearranged or grouped in different ways while retaining the same logical relationship between the inputs and the output.

For the engineer a key requirement is that the final expression is as simple as possible. As in any engineering design problem, economy is a powerful driver. Circuits that contain more components than they need, particularly if they are switch or relay contacts, are more complex, require more maintenance (in automatic telephone exchanges an ongoing maintenance task was ensuring that the thousands of relay contacts were kept clean) and, as a result, are more expensive. However, except in relatively simple cases Boolean expressions are not easy to use and manipulate. It can be difficult to see where simplifications can be made, and difficult to know whether the resulting circuit is as simple as it could be.

Although Shannon's work gave a rigorous mathematical basis to the design of switching circuits, the development of faster and more complex electronic switching circuits during the war increased pressure on engineers to find more effective ways of handling their logical equations. Work done at the Harvard Computation Laboratory (Harvard 1951) and Bell Laboratories (Keister et al. 1951) led to ways of minimising the complexity of Boolean expressions but still relied largely on tabulating all possible combinations of the variables in a chart, as well as on a good knowledge of Boolean algebra.

However, in 1952 Veitch proposed an approach, further developed by Karnaugh (Karnaugh 1953) and named for him, which shifted the problem away from mathematics towards a process of pattern recognition. Karnaugh summed up the problem facing the engineer who found themselves faced with a logic design task:

> The designer employing Boolean algebra is in possession of a list of theorems which may be used in simplifying the expression before him; but he may not know which ones to try first, or to which terms to apply them. He is thus forced to consider a very large number of alternative procedures in all but the most trivial cases. It is clear that a method which provides more insight into the structure of each problem is to be preferred. (Karnaugh 1953)

Note that Karnaugh is aiming here for 'more insight', that is he is proposing that an engineer can gain important understanding of the problem by taking an approach that does not treat the design task simply as a combinatorial exercise that involves exhaustively working through every possibility. Veitch's and Karnaugh's approach uses a grid or *map* to plot the logical value of the desired output of the circuit, either 1 or 0, against the values of the logical inputs to the circuit. What

---

[2] In this example the Boolean equation is read as: the output X will be 1 (on) if and only if (A=1 and B=1 and C=1) or (A=0 and B=1 and C=1) or (A=1 and B=0 and C=0) or (A=1 and B=1 and C=0).

were complex Boolean expressions become two-dimensional clusters of 1s and 0s. To illustrate the mapping idea, consider the Boolean equation given earlier:

$$X = ABC + \bar{A}BC + A\bar{B}\bar{C} + AB\bar{C}$$

The truth table corresponding to this Boolean expression is shown in Table 3.1(a). Table 3.1(b) shows the equivalent Karnaugh map representation.

Table 3.1(a) Truth table.

A	B	C	X
0	0	0	0
0	0	1	0
0	1	0	0
0	1	1	1
1	0	0	1
1	0	1	0
1	1	0	1
1	1	1	1

Table 3.1(b) Equivalent Karnaugh map.

	$\bar{A}\bar{B}$ (00)	$A\bar{B}$ (10)	AB (11)	$\bar{A}B$ (01)
C(1)	0	0	1	1
$\bar{C}$ (0)	0	1	1	0

The values of A, B and C are effectively used as map coordinates to identify a particular cell on the diagram in which the corresponding value of the output X is written. The three inputs produce eight possible input states[3] defining eight cells on the map, and for each state the output can be either 1 or 0. The engineer writes the appropriate value of the output in each cell and then looks for patterns of 1s to group together. In this example, the four cells containing a 1 are grouped together as shown in Table 3.1(b), with a group of two cells at the top right of the top row and a group of two cells in the centre of the bottom row. Each group of two cells is defined by just two of the three variables A, B or C: the top group is defined by BC while the bottom group is defined by $A\bar{C}$. The two groups of two cells can be written as the Boolean expression:

$$X = BC + A\bar{C}$$

[3] Since each variable can have two states, 1 or 0, three variables define $2^3 = 8$ states: 000, 001, 010, 011,…,111.

This expression is equivalent to the original equation; it is just a simpler way of writing the same logical relationship between the inputs *A*, *B* and *C* and the output *X*. But because it contains fewer terms it indicates that a simpler circuit can be built than would be the case by working directly from the original Boolean expression.

With more input variables the Karnaugh map will contain more complex patterns of 1s and 0s but the principle remains the same. Instead of using Boolean algebra the engineer marks outputs on a Karnaugh map using the inputs as the map coordinates, and then looks at the map to see which terms can be grouped together. This process of identifying and grouping terms is equivalent to applying the theorems of Boolean algebra. The engineer is effectively 'doing Boolean algebra' pictorially, allowing them to move rapidly from a specification of the logical behaviour of a circuit to a an equivalent but simplified description that can be built using fewer switches, relays or electronic logic gates.

But this is more than just a handy trick. In using this graphical method the behaviour of the circuit, from the engineer's point of view, shifts from being just a set of combinations of input and output states to clusters of output conditions that share common input states.

The engineer sees the circuit in a different way; not as a more or less random collection of physical switches or relays but more abstractly as clusters of outputs related to specific inputs. In the physical circuit the output takes only one state at a time, depending on the inputs, but in the Karnaugh map all the possible states are displayed together. Note that this is no more information than contained in the truth table, but it is represented in a different way. The map represents states as being physically near or distant from each other and this information can be used by the engineer to give an insight into problems (such as so-called *race hazards*) that may emerge when several inputs are changing rapidly together. With practice the Karnaugh map may be read to reveal behaviour that would be very difficult to deduce from the Boolean equations alone, and which can be taken into account in the design.

Although this method is practically limited in the number of input variables it can handle, its important feature is that it effectively replaces the mathematics of Boolean algebra with a simpler approach involving grouping adjacent terms on the map. Karnaugh commented:

> When inspecting the map it is not necessary to think of these subcubes [clusters of Boolean variables] by name as we must in the text, but merely observe their **relations** [emphasis added], as sets of p-squares [map areas that contain a 1]. (Karnaugh 1953, 595)

So, moving from a mathematical to a graphical representation of logical states allows the engineer to use pattern recognition to spot clusters of states that can be grouped together. With the insight provided by this representation of the system, engineering discussions of behaviour of combinational logic circuits can focus on the optimal relationships between input and output states, rather than the intricacies of Boolean algebra.

3.3 Control Systems Design

The second example of the use of visual representation in engineering is the evolution of graphical modelling techniques in frequency response analysis. A good place to start the story is with the work of Harry Nyquist (1932), Harold Black (1934) and, six years later, Hendrik Bode (1940). Although focused on the analysis and design of feedback amplifiers for communications systems their work became identified as the basis for frequency response techniques in control engineering (see the Appendix to this paper for a summary of this technique). Mindell (2002) discusses how concepts of stability and signal transmission emerged from the pre-war work of Nyquist, Black and Bode at Bell Telephone Laboratories, and how these ideas were applied more widely during World War II as engineers from different backgrounds and institutions worked on the design of gun and radar control systems.

Nyquist's 1932 paper contained a mathematical analysis of the conditions under which a feedback amplifier will be stable or unstable. Ensuring the stability of systems that contain paths where, in Nyquist's words, 'portions of the output […] are fed back to the input either unintentionally or by design' was, and remains, a significant design task. Stability is essential in any system that is intended to amplify signals or control machinery or plant. In practical terms an unstable system is usually one where feedback causes the output to oscillate wildly or vary 'out of control' (for example, causing the familiar feedback howl in audio amplification systems). In this state the output of the system is no longer determined by the input, and the system is unusable.

Nyquist used the mathematics of complex variable theory to derive his results, a treatment which many engineers, particularly those outside of communications engineering, would have found difficult to follow. Thaler (1974), in his review of classical control theory, relates anecdotally:

> …Nyquist's paper was considered as somewhat of a 'snow job' and his contemporaries would needle him about it at the various AIEE [American Institute of Electrical Engineers] meetings. Finally, over a few drinks in someone's hotel room he is said to have commented that Bell Labs had not been especially anxious to pass on the information to their competitors, so he really had not tried to simplify the paper. (Thaler 1974, 104)

If Nyquist's paper had contained mathematics alone then his results may indeed have remained safe within Bell Laboratories. However, he also included diagrams that related his mathematical results to the characteristics of a graphical plot of the frequency response of a system. He showed that by plotting frequency response data in an appropriate way, the relationship of the resulting curve to a particular point on the diagram gave information about the stability of a feedback system. Fig. 3.1(a) shows a typical Nyquist plot which uses *polar coordinates* to show the variation of a system's amplitude ratio and phase shift with frequency. In a polar plot the length of the line drawn from the centre of the diagram to the point ω_1 on the curve represents the amplitude ratio A and the angle between the line and the horizontal axis represent the phase shift φ of the system at the frequency ω_1.

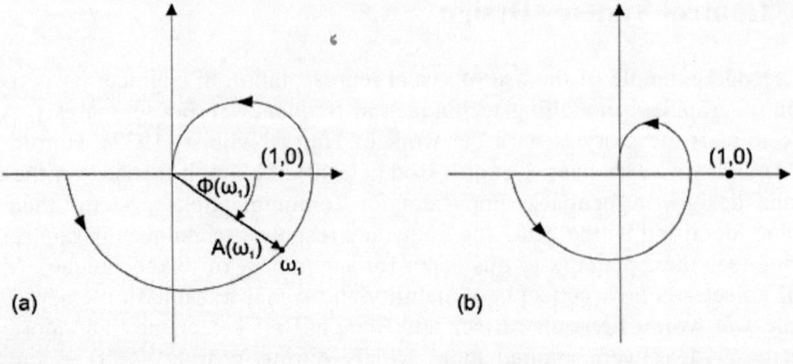

Fig. 3.1 Nyquist plots indicating the locus for (a) an unstable, and (b) a stable closed-loop system.

The locus of all the points as the frequency varies from zero to infinity (in practice this means from a relatively low to a relatively high frequency) gives the curve on the Nyquist plot.

Even if they were not able to follow the formal mathematics, engineers found the Nyquist plot straightforward to use. Nyquist established simple rules based on whether the point (1, 0) lay 'inside' or 'outside' the curve. By plotting the frequency response locus engineers could tell at a glance whether their system would be unstable, as in Fig. 3.1(a), or stable, as in Fig. 3.1(b). Nyquist's own words are revealing:

> The circuit is stable if the point lies wholly outside the locus x = 0. It is unstable if the point is within the curve. It can also be shown that if the point is on the curve then conditions are unstable. We may now enunciate the following rule: Plot plus and minus the imaginary part of AJ(iω) against the real part for all frequencies from 0 to ∞. If the point 1 + i0 lies completely outside the curve the system is stable[4]. (Nyquist 1932)

The significant point here is that, even at this early stage, Nyquist's language has moved away from explaining and predicting the behaviour of the electrical circuit he is considering in terms of its voltages, currents and physical components. His comments are wholly in terms of the relationship between the frequency response locus, which is simply a line drawn on paper, and a particular point on the diagram. From a design perspective, where a system has not yet been built or may not yet be working correctly, Nyquist's language is veering towards a causal explanation of behaviour: *if* the curve and the point are in the right relationship to each other *then* the circuit or system will be stable. In other words, when engineers use graphical models like the Nyquist plot they don't need to know the physical details of a particular system or much about the formal mathematics used to model it in order to have a meaningful conversation about key characteristics of its

[4] $AJ(i\omega)$ is the open loop frequency response function of the system. It is a complex function of the frequency ω and may be represented equivalently by its real and imaginary parts, or by its amplitude and phase as described above.

dynamic behaviour[5]. However, the language of representation and interpretation takes time to learn and become established. In their work on feedback amplifiers both Black (1934) and Bode (1940) reiterated Nyquist's stability criterion but subtly modified the language they used. In his paper Black, considering the problem of how to avoid 'singing' (unstable oscillations) in amplifiers, wrote:

> If the resulting loop or loops do not enclose the point (1, 0) the system will be stable, otherwise not.

In a footnote, as if aware that diagrams do not speak for themselves and new ways of talking about them are not inherently obvious, he refers readers back to Nyquist's paper for a description of 'exactly what is meant by enclosing the point (1, 0)'.

By the time Bode is writing in 1940 he is talking about Nyquist curves that 'encircle' or 'enclose' a point (neither term was used originally by Nyquist) without further explanation or justification. For technical reasons he also re-presents Nyquist's original diagram by rotating it through 180 degrees 'so that the critical point occurs at -1, 0 rather than +1, 0'. In doing so Bode established the form of the Nyquist diagram, as well as the language of enclosure or encirclement, still used today to determine the stability of closed loop control systems.

Nyquist's diagram was not the only way of plotting and interpreting frequency response data. Black and Bode both offered other ways that were to become significant in the graphical control system design methods that emerged in the 1940s and 50s (Bennett 1993). Black used rectangular coordinates to plot a system's amplitude ratio against phase shift for a set of frequency points, laying the foundations for a design tool later known as a Nichols chart (James, Nichols and Phillips, 1947, 219).This representation not only allowed engineers to plot open-loop frequency response data and determine the stability of a system when the feedback loop was closed, but also gave information about how close the system was to instability. Bode's contribution was to define more clearly these 'margins of stability' and, using plots of amplitude ratio and phase shift against frequency on logarithmic coordinates, to introduce approximations that allowed complex transfer functions to be easily plotted as straight lines. Together the work of Nyquist, Black and Bode, although applied to feedback amplifiers rather than to control systems as such, helped transform the mathematics of control system design into a number of graphical processes.

By the early 1940s frequency response techniques were well known among communications engineers in the USA and, owing to the demands of the war, were also becoming known to a new generation of control engineers concerned with the design of accurate position and velocity control servomechanisms. Oldenburger (1954b) writing in a Foreward to papers presented at a frequency response symposium held by the American Society of Mechanical Engineers (ASME) illustrated how engineering communities were being brought together:

---

[5] For a more detailed account of this linguistic shift see (Bissell and Dillon 2000) and (Bissell 2004).

> … it became evident practically simultaneously to both a group under the direction of Prof. Gordon Brown at the Massachusetts Institute of Technology, and a group at the Bell Telephone Laboratories, that the contributions of Nyquist, Bode, and Black could be applied advantageously to automatic-control problems as well as those of communication. This interest in frequency-response techniques was precipitated in part by the rigid military requirements for aiming antiaircraft guns by radar. (Oldenburger 1954b, 1145)

Oldenburg's comments applied to American engineers but similar realizations occurred more or less independently elsewhere, although they were not always so well developed. While this chapter focuses on English-language control texts of the 1940s and 50s primarily from the USA and the UK, there were also pre-war and post-war classic works published in Germany, France, Italy and the former Soviet Union. For a more detailed account and extensive references see Bissell (1999).

It is no surprise that servomechanism designers largely adopted the language and techniques of communications engineers, and became quite comfortable in describing their control systems in the same terms. Communications terminology such as bandwidth, gain (measured in decibels), phase shift, and gain and phase margins, together with a range of ways of plotting frequency response data, became the natural way of talking about the behaviour of these electromechanical gun-aiming and tracking systems. Here is Enoch Ferrell of Bell Telephone Laboratories writing about the design of servomechanisms in1945:

> The servo signal is usually mechanical; the circuit elements are motors, gears, or thermostats. The noise and distortion are called error. But the basic problems of stability, bandwidth and linearity are just the same. And the simple, straightforward design technique that is based on circuit theory, and that has been used so successfully with negative feedback-amplifiers, is just as simple and straightforward when used with servo systems. (Ferrell 1945, 763)

In his description of part of the design process for a typical servo system he refers to a plot of the system's frequency response characteristics. A version of this plot is shown in Fig. 3.2. It is worth including the extract at some length because it gives a flavour of the type of explanation engineers would be familiar with.

> To reduce errors we can increase the gain […] in the vacuum-tube amplifier. But this will destroy the margins and cause singing at some higher frequency. Can we introduce a phase-correcting network that will improve the phase margin and permit the use of more gain? Yes, we can. And, by analogy to amplifier and wire transmission practice we will call that network an equalizer… [The plot shows] the gain and phase of a servo loop [before equalization with Gain 1 and Phase 1 and] after equalization [with Gain 2 and Phase 2]. Here loss was introduced just below the 6-to-12 corner of the original curve. This was done by means of a shunt capacitor in a direct-current part of the system. It gave a steeper slope and more phase shift. Then a resistance was put in series with the capacitance, so that we would get a flat loss at higher frequencies and recover the lower phase shift… [I]n the region of gain crossover [where the gain is 0dB] we have a moderate slope in the gain curve and hence a safe phase margin. But at a little lower frequency we have a steep section. This has raised the low-frequency end of the curve [so that Gain 2 is higher than Gain 1], thereby giving us a large low-frequency μ and hence small errors. (Ferrell 1945, 767)

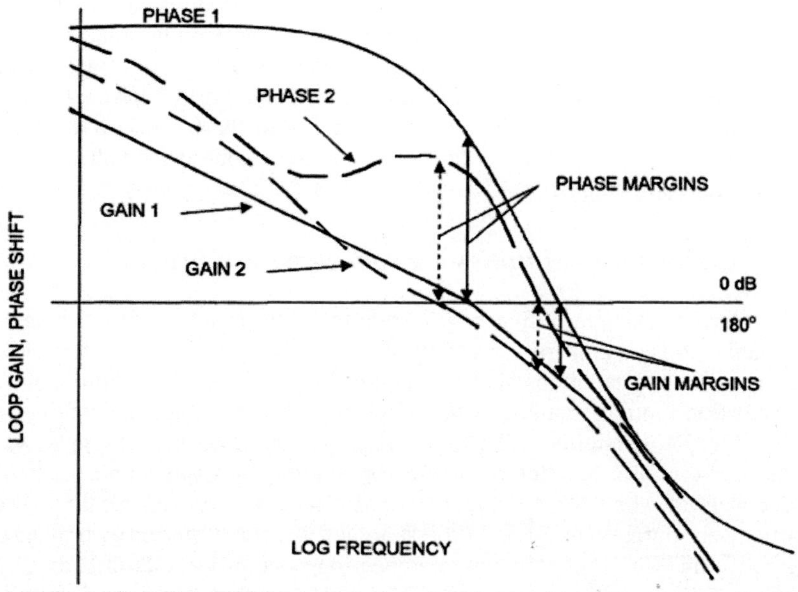

Fig. 3.2 Frequency response plot. Redrawn by the author after Ferrell (1945).

In this mix of technical jargon (errors, gain, phase, singing, loss, equalization, low and high frequency, μ), features of the frequency response plot (6-to-12 corner, slope, phase margin, crossover, steep section) as well as references to real components (vacuum-tube amplifier, capacitor, resistance), we might be listening to the engineers talking through the design as they build a shared understanding of system behaviour.

Note also that the frequency response plots Ferrell is referring to are little more than sketches. The axes are not scaled and indicate only the quantity being plotted: logarithmic frequency along the horizontal axis, and loop gain (in decibels) and phase shift (in degrees) along the vertical axis. The only numerical indication of gain and phase shift is that the loop gain is 0 dB and the phase shift is 180 degrees where the relevant curves cross the horizontal axis. The plot contains four curves which show the loop gain and phase shift of the system before (Gain 1 and Phase 1) and after it has been 'equalized' (Gain 2 and Phase 2, shown as dashed lines) to improve performance. Despite the apparent lack of detail, however, such a sketch contains all the important features of the system from a design engineer's point of view. As Ferrell's account indicates, the emphasis is on qualitative features rather than detailed numbers, which can always be worked out later once the general form of the system's behaviour has been established.

What Ferrell's description reveals is the richness of this deceptively simple representation that allows the engineer to understand key features about system stability and behaviour both at high frequencies (indicating fast-acting performance) and lower frequencies (indicating slower responses and steady state errors). And, further, it illustrates how engineering explanations of systems behaviour draw on a complex mix of physical understanding and graphical features conventions to generate an acceptable explanatory account. So, for example, in the final sentence the desired '...large low-frequency μ [gain] and hence small errors' are apparently produced by the '...raised low-frequency end of the curve...'. In this way of talking the graphical representation and manipulation of the frequency response data has become as real and as influential on system behaviour as the voltages, currents and physical components.

3.3.1 Divided Communities – From Servos to Processes

While servomechanism design engineers seemed easily to adopt the language of communications engineering, process engineers concerned with the control of large industrial installations such as chemical and power generation plant, found the transition more difficult. For them not only was the language coming from a quite different discipline of engineering but the systems concepts such as frequency response, transfer functions and stability margins, concepts that were independent of whether the component was electrical, mechanical, thermal or hydraulic, were unfamiliar. Bissell (2004) argues that these tensions were the source of creative thinking about the ways engineering systems were described:

> ...the accounts at conferences and learned society presentations testify to the rather difficult process in adopting the 'systems' ways of thinking and the novelty of applying general systems ideas outside their 'natural' home of communications and electronic engineering. It was not a question of simply applying existing communications theory and techniques to other systems [...] Rather, the collaborators from different technical backgrounds renegotiated their modelling techniques, and the language in which they were expressed, so as to abstract the essential. (Bissell 2004, 323)

However, in the fragmented control engineering community after the war the collaboration and renegotiation of ways of thinking that had served some groups in good stead would not be achieved overnight. While the servo engineers had made significant progress with gun and radar servomechanisms there had not been the pressure on process engineers to change in the same way. Given that the lifetime of process plant is measured in decades, emphasis had been on maintaining steady and reliable output throughout the war years. In many cases this could be achieved as well by human operators as by automatic control. Indeed, as late as 1955 A.J. Young, Head of the Central Instrument Laboratory at Imperial Chemical Industries (ICI) Ltd. in the UK, warned that, although attitudes were changing,

> There still lingers a feeling among management in certain industries, or branches or industry, that automatic control is a luxury which can well be dispensed with in the majority of plants, although it must be regarded as an unfortunate necessity in some. (Young 1955, 9)

The development of automatic process control systems during and after the war, therefore, proceeded cautiously. An internal report written by C.I. Rutherford, a process control engineer in Young's laboratory at ICI, around 1949 (the report is not dated), included an extensive bibliography of some 130 references going back to the early 1930s which 'had any bearing on process control'.

However, clearly Rutherford was not impressed with what he had found, and noted a 'wide gap' between theoretical treatments and practical papers:

> It is not felt that the majority of the articles listed will provide any useful reading for the average Process Control engineer, but it was felt worthwhile including everything that had been seen, even if only as a warning to others not to waste time searching for articles and reading them. (Rutherford c1949, 2).

One reason for the difficulty was that process engineers came from a quite different engineering tradition. While the servo designers had backgrounds in light current and communications engineering, and so would have been aware of at least some frequency response ideas as part of their training, the process engineers were more likely to have had backgrounds in mechanical or heavy electrical engineering. Although they would have been familiar with mathematical models and differential equations, it was the 'systems approach' of the communications engineers, that is, a focus on input/output behaviour of operational 'black boxes' such as filters, amplifiers, equalisers, transducers and controllers rather than on individual components, that pointed the way to modern design (Bissell 2004). Further, their perceptions about what constituted effective control were different. For the servo engineer, the aim was to design fast-acting mechanisms that could track moving objects accurately. Response times were measured in fractions of a second, tracking errors had to be small and the system had to respond rapidly to essentially randomly changing inputs, all in all a very similar set of requirement to those for a communications system. In this context, talk of high bandwidth and frequencies that extended to tens or hundreds of Hertz (cycles per second) was perfectly normal.

For the process engineer, however, the norms were quite different. Control of large industrial plant usually meant keeping the plant processes at a steady state in the face of external disturbances, for example keeping steam temperatures and pressures steady in a power plant as electrical demand changed, or maintaining steady flow rates over long periods in a distillation process. When changes did occur a large industrial plant might take hours, sometimes days, to settle down again to steady behaviour. Talk of frequency, bandwidth and gain measured in decibels seemed unnatural and unhelpful as Ed Smith, an American process control engineer (citing the 1945 paper by Ferrell), pointed out

> The experts in the servomechanism and control fields are unfortunately not exceptions to the rule that the experts in an art always ball up the terminology and notation until they can be followed only by a like expert [...] At least the controller [process engineer] can directly express a ratio of effect to cause [...] without becoming involved in the width of a hypothetical frequency transmission band and in such acoustical abstraction as 'decibels per octave' [...] A noteworthy exception to this is that grand tool the Nyquist method of testing stability. This is essentially so simple that it is difficult even for experts to confuse although they bring in the 'complex conjugate' and the idea of 'negative frequencies'. Discussion of Brown and Hall (1946, 523)

Although process engineers in the UK in the late 1940s and early 50s might well have echoed Smith's words, many of the changes that had taken place on the other side of the Atlantic had yet to filter across. The British engineer G.H. Farringdon, for example, included a somewhat oblique discussion of frequency response and included Nyquist-like polar plots in his 1951 treatment of automatic control, but did not explicitly address or reference Nyquist's stability criterion (Farrington 1951). It seems that, although listed in Rutherford's ICI control bibliography (interestingly neither Bode nor Black are included), Nyquist's ideas were still regarded with some unease:

> The method of frequency response analysis is explained as a simple and logical procedure, and its use in conjunction with Frequency Response Diagrams is shown to be easily intelligible without the use of any involved mathematics; such mathematical phenomena as complex variables and polar co-ordinates, which normally adorn the literature on frequency response, have been carefully avoided, in order that those unfamiliar with such techniques will not feel that the science of automatic control is still one for the specialist and return once more to the method of trial and error. (Rutherford c1949, 20)

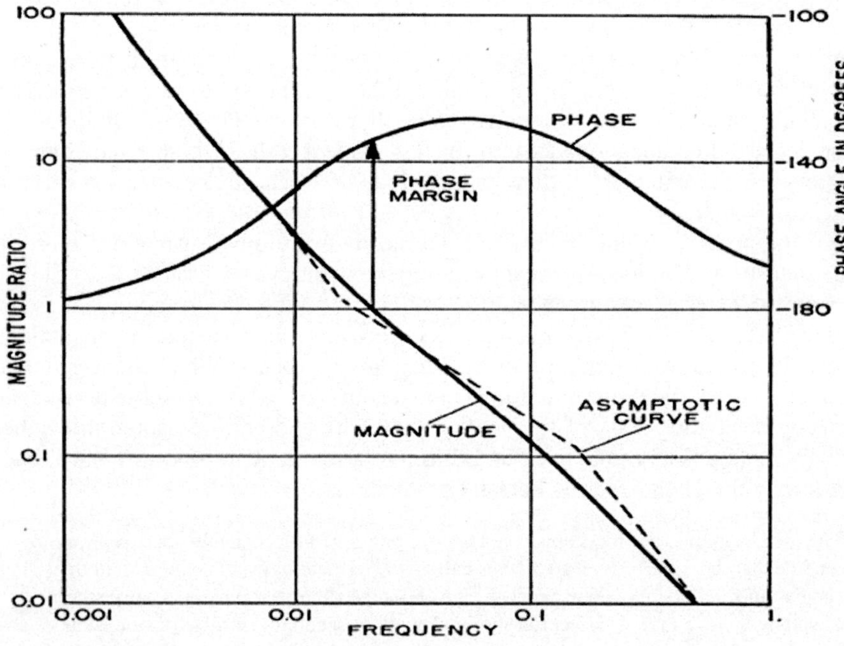

Fig. 3.3 Frequency response plot using the US convention. Source: Oldenburger (1954, 1159). Reproduced by permission of the American Society of Mechanical Engineers.

For the engineers at Imperial Chemical Industries, at least, their approach to frequency response representation had evolved along rather different lines to their American counterparts, and these differences started to come to a head as engineers began to meet again after the war at international conferences.

In 1954 the American Society of Mechanical Engineers organised an international symposium to share practical and theoretical developments in frequency response methods, and to propose a standard approach for presenting of frequency response data. Figs. 3.3 and 3.4 show examples of frequency response plots taken from two papers at the symposium. Fig. 3.3 shows a typical plot (Oldenburger 1954, 1159) drawn using the emerging American convention, while Fig. 3.4 (Aikman 1954, 1317) uses the style favoured by the UK engineers at ICI. While superficially the same as the gain and phase plots used by American engineers, the ICI data is plotted in a quite different way. Both curves contain amplitude and phase information but in Fig. 3.3 this is plotted against increasing frequency on the horizontal axis, while in Fig. 3.4 it is plotted against increasing period of oscillation, which, as the annotation *frequency in cycles per second* at the top of the graph shows, is the inverse of frequency.

Similarly, the American convention was to plot the magnitude, or amplitude, ratio (the ratio of the amplitude of the steady state output sinusoidal variation to the amplitude of the input sinusoidal variation) along the vertical axis, sometimes expressed in decibels[6], while the ICI convention was to plot 'attenuation', which is the inverse of the magnitude ratio. The implication of this is clear; engineers used to assessing the design and performance of a system using the shapes and slopes of the curves produced in one convention, would not be able to easily interpret the different curves produced by the other. Writing in 1950 Rutherford offers a rationale for the ICI approach:

> …the phase differences and amplitude ratios of the applied and resulting oscillations are plotted against the period of oscillation. Period of oscillation is chosen in preference to frequency because it is more easily visualised and appears in units of time, which are generally familiar; the period scale is logarithmic in order to include a reasonable range in a small space. Against the period of oscillation is plotted the phase difference, using a linear scale in degrees on the upper half of the frequency response diagram, and the amplitude ratio on a logarithmic scale on the lower half. An alternative would be the use of a decibel scale, but this brought an unnecessary concept which might not be familiar to some of those who may wish to apply the treatment. Frequency response diagrams are therefore plotted on composite paper having two vertical scales. (Rutherford 1950, 335)

However, this approach was under attack. 'Except in England…' wrote Oldenburger (1954a) in his symposium paper recommending standards for the presentation of frequency response data, '…the frequency f is used in preference to the period P as the co-ordinate on the horizontal axis […] The use of decibels is standard in the Bell Laboratories approach'. His comments clearly struck a chord with his audience. Arnold Tustin, then professor at the University of Birmingham, UK, and one of the early pioneers of sampled-data control systems[7], commented:

---

[6] In control engineering an amplitude ratio A is commonly expressed in decibels as $20\log_{10} A$.

[7] In an interview shortly before his death Arnold Tustin talked to Chris Bissell about his contributions to control engineering during the Second World War (Bissell 1992).

There is certainly a need for some such recommendations [...] One notes, for example, that our friends from ICI [...] still insist on being the odd man out, by plotting everything inside out and upside down, to the general confusion. It is to be hoped that the present discussion will bring differences in practices of this kind to an end. (Discussion of Oldenburger (1954a))

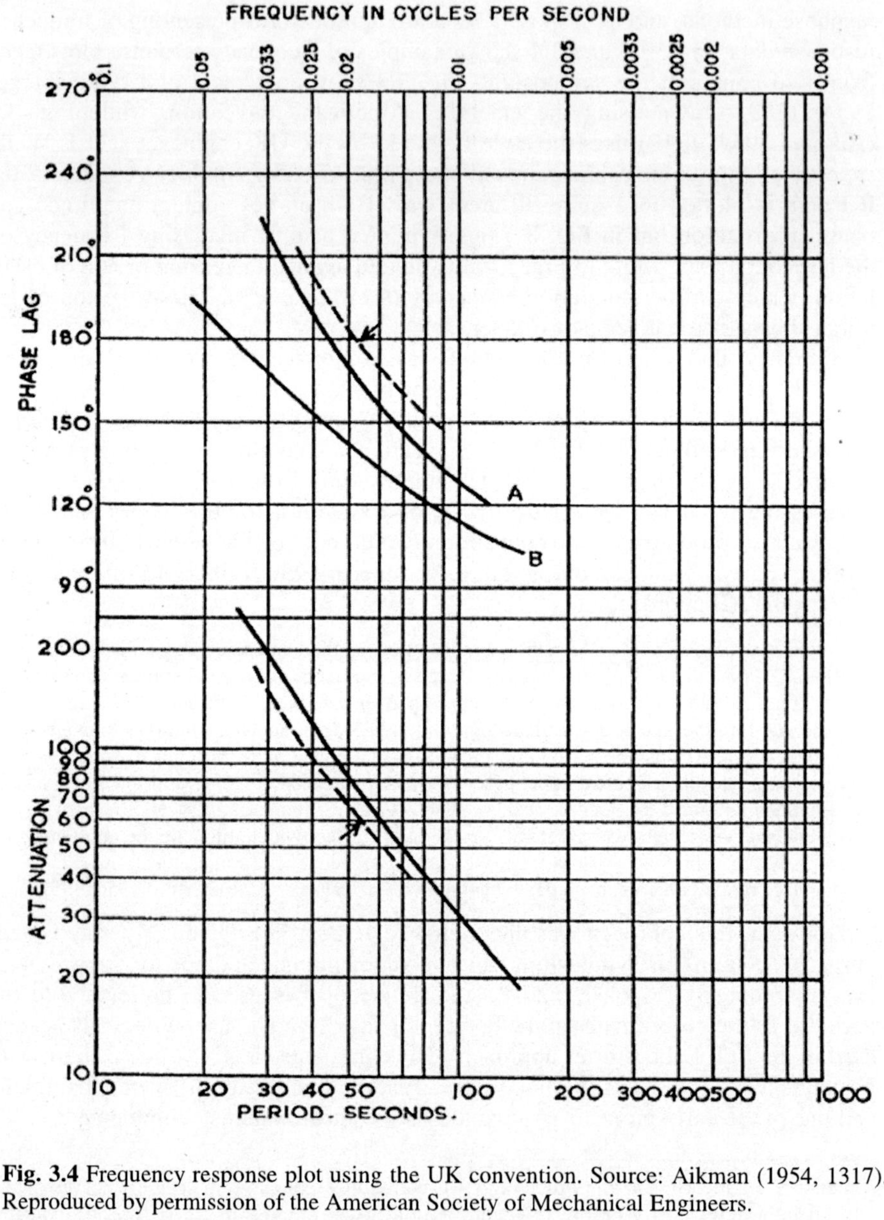

Fig. 3.4 Frequency response plot using the UK convention. Source: Aikman (1954, 1317). Reproduced by permission of the American Society of Mechanical Engineers.

And even 'our friends from ICI' seemed to agree. In the discussion following his own paper at the same symposium, A.R.Aikman, one of Rutherford's colleagues from ICI's Central Instrument Section, remarked:

> The author [has also] been struck by the differences between European and American terminology. It is certainly hoped that a larger area of agreement between rival terminologies can be reached in the near future, and that the present anomalous position, in which there are different terms in the same language for the same basic concepts, will be rectified. The graphical representation used in this paper is indicative only of the practice in the author's company and is not to be taken as representative of the European practice as a whole; if terminology could be unified the interpretation of graphs would present little difficulty[8]. (Aikman 1954, 1322)

From a community of practice standpoint representing systems in different ways presented few problems so long as practices remained local. What geographically-separated groups decided to do had relatively little impact on each other. In the emerging control engineering community freed from the restrictions of wartime, however, it became clear that different styles and modes of representing dynamic system information were getting in the way of further development. Engineers used to recognising at a glance the features of a particular system were not able to easily interpret the diagrams and the associated explanations when they came from outside their local community. So long as the pictorial forms were different then the ability to construct a common language of practice, such as Ferrell used to talk through his design, was severely limited.

Practice varied across Europe too, but by 1952 was already moving towards common standards (Bissell 1994). Given the emphasis on the importance of graphical methods in understanding and designing systems behaviour, a common interpretation was clearly essential if different groups of engineers were not to spend time unravelling each others' notation and diagrams. The skill of the engineer in understanding system behaviour implied that the shape of a curve could be read at a glance. Recognising the key points in a frequency response plot and telling a good story about them, as Ferrell demonstrated, was part of the tradecraft of engineering, and a shared tacit skill within the community. An engineer who couldn't easily read and interpret the diagrams would simply, in Wenger's words, not have the 'experiences, stories or tools' that would allow them to communicate or participate effectively.

A final comment: a popular conception is that the war produced a step change in thinking in applications of new technologies and techniques that came into effect as soon as the war was over. Bissell observes:

> … almost every early postwar paper made some reference to a 'new language', to 'problems with terminology', to the need to 'translate' the 'jargon' of one or other group. But once agreement had been reached on language and conventions, the conceptual difficulties […] appeared to fade away. Conventions, once agreed, become invisible to the insider … (Bissell 1994)

[8] It is interesting to note that despite this wish for a unified approach A.J.Young, the head of the group at ICI to which both Aikman and Rutherford belonged, stuck to the 'inside out and upside down' style of plotting frequency responses in his own classic book on process control published in 1955 (Young 1955).

However, not all such changes impacted immediately. In this example we see major shifts in thinking amongst communication and servo engineers, but a much slower take-up amongst process engineers, who took 10 years or more to come terms with new techniques, new ways of thinking and in particular, new ways of talking. In the UK it would take until the mid to late 1950s before the new techniques and models become part of the everyday assumed understanding of process control engineers.

3.4 Concluding Remarks

The discussion in this chapter centres on the idea that engineers have developed visual or pictorial ways of representing systems that not only avoid the use of complex mathematics, but have enabled ways of seeing and talking about systems that draw on the graphical features of the models. I have suggested that this approach develops within a community of engineering practice where the interpretation and understanding of these visual representations of systems behaviour are learnt, shared and become part of the normal way of talking.

The Karnaugh map, for example, is a simple illustration of how switching engineers were able to replace Boolean algebra and design efficient circuits by a simple visual pattern recognition technique. New shared ways of seeing and understanding within the community make the mathematical details of the model effectively invisible; such details do not have to be explained (except to novices) in engineering conversations.

The more complex example of the development of graphical control engineering design methods illustrates that the spread and take-up of such techniques across the related communities of practice of servo and process control engineering after the war involved reconciling differences in language and graphical styles that had evolved in separate groups.

For different groups of engineers concerned with aspects of system design but separated by wartime restrictions, technical cultures, and geographical location, the development of these techniques occurred at different rates and with variations that made cross-community communication difficult. Perhaps unsurprisingly, therefore, control engineers in different industries did not all emerge after the war sharing ideas that had been developed under wartime secrecy. Rather the models and language that engineers used to visualise and talk about system behaviour all took time to be widely accepted and understood across the community.

In his book *What engineers know and how they know it* Vincenti (1990) comments on the 'messiness' of the evolution of engineering concepts and practices and the difficulty, perhaps the impossibility, of telling a single simple story about them:

> As any engineer knows, the technological learning process always requires more effort in fact than appears necessary in hindsight. [...].The learning, in short, while it is going on is messy, repetitious and uneconomical. It is easy in retrospect to read more logic and structure into the process than was most likely evident to those concerned. (Vincenti 1990, 11)

Evolutionary messiness notwithstanding, what is significant in the examples considered in this paper is that a graphical approach to design appears not just as a convenient pictorial form but as a way of gaining deep insight into the relationship between systems variables and ways that a design can be improved. Oldenburg sums up the power of this approach by conjuring the vision of system response curves that, for communications and control engineers, are unique and revealing representations of physical behaviour:

> Except for errors in measurement, the experimental frequency response curves for a physical device truly represent this device. No two physical systems have the same frequency response curves. Such curves are in a sense as characteristic of a physical system as fingerprints are of a human being. They are really differential equations in pictorial form. (Oldenburger 1954a)

Design engineers become skilled at interpreting and using graphical or pictorial models by 'reading in' to the model their experience and knowledge to generate an explanation about system behaviour, manipulating the shape of the graphical model guided by the desired system characteristics, and then 'reading out' from the model to generate a procedure for defining or changing the behaviour of the real system. The engineer 'puts the model to work', to use Morgan and Morrison's phrase, by using it as the focal point for a story or conversation about how the system behaves and how that behaviour might be changed. It is by mediating in this process by focusing language to stress some features of the real system while ignoring others that models contribute to new shared understandings in a community of engineering practice.

Appendix

In the frequency response method a control system is characterised by how the output of the system responds to sinusoidally varying input signals over a range of frequencies. In control engineering terms an output may, for example, be a position or velocity in a servomechanism, or a temperature, pressure or liquid level in a process plant,. The corresponding input is often the desired value (or set point) of the output. If the system is linear (or its actual behaviour can be adequately modelled as a linear system over the range of operation) then, when the system has settled to a steady sinusoidal response, the output variation of the system will be a sinewave when the input is a sinewave, and will vary at the same frequency as the input.

At any particular frequency the input and output sinewaves will in general have different amplitudes, and the peaks and troughs of the output variation will not occur at the same time as those of the input. The output sinewave is said to be *shifted in phase* relative to the input sinewave. The *frequency response* of the system at the frequency ω is defined by two numbers: the ratio $A(\omega) = X_o/X_i$ of the amplitude of the output sinewave X_o to the amplitude of the input sinewave X_i, and the phase shift $\varphi(\omega)$ which is the amount by which the output is shifted in phase relative to the input. The variation of $A(\omega)$ and $\varphi(\omega)$ with the frequency ω can be plotted in different ways to give the engineer insight into the behaviour of the system

References

Aikman, A.R.: Frequency-response analysis and controllability of a chemical plant. Transactions of the American Society of Mechanical Engineers 76(8), 1313–1323 (1954)

Bennett, S.: A History of Control Engineering 1930-1955. Peter Peregrinus, Stevenage (1993)

Bissell, C.C.: Pioneers of control: an interview with Arnold Tustin. Institution of Electrical Engineers Review 38(6), 223–226 (1992)

Bissell, C.C.: Spreading the word: aspects of the evolution of the language of measurement and control. Measurement and Control 27(5), 149–155 (1994)

Bissell, C.C.: Models and "black boxes": Mathematics as an enabling technology in the history of communications and control engineering. Revue d'Histoire des Sciences 57(2), 307–340 (2004)

Bissell, C.C., Dillon, C.R.: Telling tales: models, stories and meanings. For the Learning of Mathematics 20(3), 3–11 (2000)

Black, H.S.: Stabilized feedback amplifiers. Bell System Technical Journal 13(1), 1–18 (1934)

Bode, H.W.: Relations between attenuation and phase in feedback amplifier design. Bell System Technical Journal 19(3), 421–454 (1940)

Brown, G.S., Hall, A.C.: Dynamic behaviour and design of servomechanisms. Transactions of the American Society of Mechanical Engineers 68, 503–524 (1946)

Farrington, G.H.: Fundamentals of Automatic Control. Chapman and Hall, London (1951)

Ferguson, E.S.: Engineering and the Mind's Eye. MIT Press, Cambridge (1992)

Ferrell, E.B.: The servo problem as a transmission problem. Proceedings of the Institute of Radio Engineers 33(11), 763–767 (1945)

Harvard: Synthesis of Electronic Computing and Control Circuits. Harvard University Press, Cambridge (1951)

James, H.J., Nichols, N.B., Phillips, R.S.: Theory of Servomechanisms. McGraw-Hill, New York (1947)

Karnaugh, M.: The map method for synthesis of combinational logic circuits. Transactions of the American Institute of Electrical Engineers Pt 1 72(9), 593–598 (1953)

Keister, W., Richie, A.E., Washburn, S.H.: The Design of Switching Circuits. Van Nostrand, New York (1951)

Layton, E.T.: American Ideologies of Science and Engineering. Technology and Culture 17(4), 688–701 (1976)

Mindell, D.A.: Between Human and Machine. Johns Hopkins University Press, Baltimore (2002)

Morgan, M.S., Morrison, M.: Models as mediating instruments. In: Morgan, M.S., Morrison, M. (eds.) Models as Mediators, pp. 10–37. Cambridge University Press, Cambridge (1999)

Nyquist, H.: Regeneration Theory. Bell System Technical Journal 11, 126–142 (1932)

Oldenburger, R.: Frequency-response data, standards and design criteria. Transactions of the American Society of Mechanical Engineers 76(3), 1155–1169 (1954a)

Oldenburger, R.: Frequency Response Symposium - Foreword. Transactions of the American Society of Mechanical Engineers 76(3), 1145–1149 (1954b)

Rutherford, C.I.: The practical application of frequency response analysis to automatic process control. Proceedings of the Institution of Mechanical Engineers 162(3), 334–354 (1950)

Rutherford, C.I.: Automatic Control Theory and Application. Imperial Chemical Industries Ltd., Technical Dept., Central Instrument Section, Internal Report No. 480292/HO/NH/TC (c.1949)
Shannon, C.E.: A symbolic analysis of relay and switching circuits. Transactions of the American Insitute of Electrical Engineers 57, 713–723 (1938)
Shannon, C.E.: A mathematical theory of communication. Bell System Technical Journal 27(3), 623–656 and 27(4), 623–656 (1948)
Smith, E.S.: Automatic Control Engineering. McGraw-Hill, New York (1944)
Thaler, G.J. (ed.): Automatic Control: Classical Linear Theory. Dowden, Hutchinson and Ross, Stroudsburg, PA (1974)
Vincenti, W.: What Engineers Know and How They Know It. Johns Hopkins University Press, Baltimore (1990)
Wenger, E.: Communities of Practice. Cambridge University Press, Cambridge (1998)
Young, A.J.: An Introduction to Process Control. Longmans, London (1955)

Chapter 4
Metatools for Information Engineering Design*

Chris Bissell

The Open University, UK

Abstract. An examination of the professional practice of engineers in many disciplines reveals a history of engineers developing highly sophisticated tools to eliminate the need to 'do mathematics' in the conventional sense. This chapter will build upon the previous one by Dillon to consider further aspects of the history of a number of what I shall call mathematical 'meta-tools' in the fields of electronics, telecommunications and control engineering. In common with Dillon I argue that, for most engineers, 'doing mathematics' has become something categorically different from the mathematics of physical scientists or mathematicians. The chapter concentrates on the origins and changing fortunes of a number of classic information engineering meta-tools that appeared in the period just before or after the Second World War: Bode plots (late 1930s); the Smith chart (1939); the Nichols chart (1947); phasor, spectral and signal constellation models (throughout the period); and the root-locus technique (1948). The 1950s and 1960s saw an increasing mathematicisation of engineering education, linked to the rise of the notion of 'engineering science' that was driven to a large extent by the legacy of WW2 research and development and the post-war funding environment in the USA and elsewhere. Such changes, and the arrival of digital computers, meant that the utility of the earlier diagrammatic tools was often played down or questioned. In recent years, however, such tools have been incorporated into powerful engineering software, where their function now is not to avoid computation, but to mediate between the user and the machine carrying out the computation.

4.1 Introduction

Historians and analysts of electronics and telecommunications have tended to concentrate on enabling technologies such as the vacuum tube, the transistor, the

* Earlier versions of parts of this chapter appeared as (1) Mathematical 'meta tools' in 20th century information engineering, in *Hevelius*, 2, 2004, pp. 11-21 and (2) Models and "black boxes": Mathematics as an enabling technology in the history of communications and control engineering, in *Revue d'histoire des sciences*, 57(2) juillet-décembre 2004, pp. 305-338. (Bissell 2004a, b).

C. Bissell and C. Dillon (Eds.): Ways of Thinking, Ways of Seeing, ACES 1, pp. 71–94.
springerlink.com

microprocessor, and so on – or on the socio-political aspects of developing, regulating, and managing large-scale communication systems. Yet just as significant – but largely neglected by historians and analysts of technology – is an approach to modelling, analysis and design based on a quintessentially 'communications engineering' use of mathematics. This approach, ultimately characterised by terms such as 'linear systems theory' and 'black box analysis' is still a key factor in the development of communications devices and systems, and highly influential in other areas of engineering.

This chapter builds on the previous one by addressing some further aspects of the development of mathematical modelling in communications and control engineering, identifying a number of crucial features of such modelling, and setting them, albeit briefly, in the wider context of communications and control engineering technology, including instrumentation, computation, and simulation. The treatment is necessarily fairly terse; for extended background the reader is referred to the classic texts by Bennett (1993) and Mindell (2002) as well as to the series of papers by Bissell (1986–2009). These sources also include detailed references to original historical papers that are not always included here.

4.2 Phasors, Filters and Circuits

Models of electronic devices, and the explanations of the propagation of electromagnetic waves, rely on complex numbers and complex variable theory. From the early days, engineers came up with radical new graphical techniques to aid their understanding, and to remove the need for calculations that were difficult in the days of slide rules and mechanical calculating machines.

4.2.1 Phasors

One of the earliest of such tools was the use of phasors, a vector representation of complex numbers. Essentially, in the phasor approach a sinusoid is characterised by a complex number[1] (phasor) $\boldsymbol{V} = A \exp(\mathbf{j}\phi)$, such that $|\boldsymbol{V}| = A$ represents the amplitude of the sinusoid and Arg $\boldsymbol{V} = \phi$ represents the relative phase defined by the deviation from an 'in-phase' reference. An ideal linear electrical component can thus be characterised by a complex gain $\boldsymbol{G}$, so that $|\boldsymbol{G}|$ represents the ratio of the amplitude of the output sinusoid to that of the input, and Arg $\boldsymbol{G}$ represents the phase shift introduced by the component. Charles Steinmetz, one of the pioneers of this field wrote in 1883:

> The method of calculation is considerably simplified. Whereas before we had to deal with periodic functions of an independent variable 'time', now we obtain a solution through the simple addition, subtraction, etc of constant numbers … Neither are we restricted to sine waves, since we can construct a general periodic function out of its sine wave components. (Steinmetz 1883, 597, trans.)

[1] I shall use the electrical engineering convention of **j** rather than **i** for √-1.

> … With the aid of Ohm's Law in its complex form any circuit or network of circuits can be analysed in the same way, and just as easily, as for direct current, provided only that all the variables are allowed to take on complex values. (Steinmetz 1883, 631, trans.)

Incidentally, of this paper, Steinmetz later wrote: '... there was no money to publish the Congress paper, and the paper remained unpublished for years, and the symbolic method unknown'. Yet the paper did appear in German the same year as the Congress, and it is from this source that the above citation is taken (author's translation).

The phasor approach allowed powerful geometrical representations (based on Argand diagrams) to be used for a wide variety of applications: from electrical power transmission to electronic circuits and electromagnetic wave transmission. Engineers became used to manipulating these diagrams without recourse to the complex number representations that many of them found difficult to understand. For example, the phasor diagram of a sinusoid affected by noise can be represented geometrically by a vector representing the amplitude and phase – that of the transmitted 'pure sinusoid' plus a randomly varying noise vector which introduces a degree of uncertainty into the resultant received phasor. In the late twentieth century, this general idea was transformed into the 'signal constellation' model of digital signals, of which a little more later.

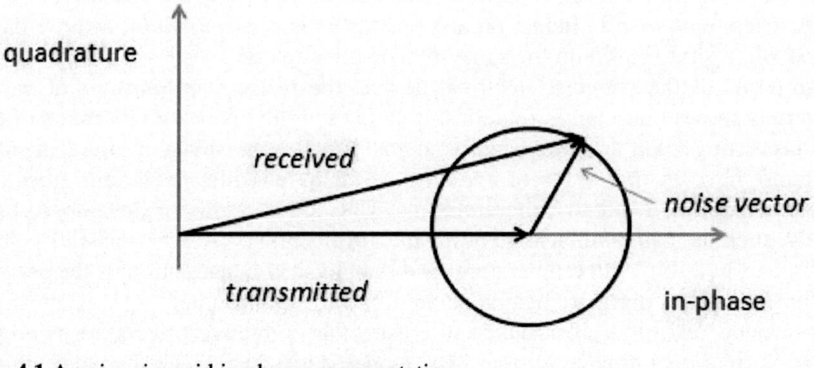

Fig. 4.1 A noisy sinusoid in phasor representation.

The phasor approach was the first step towards powerful frequency-response models of electrical and electronic devices. By the 1920s much of the theory of the synthesis of filters (required for telegraphy and telephony) had been developed, and again, graphical techniques were used rather than traditional mathematical approaches. Filter designers learned how to manipulate circuit diagrams and frequency response plots in order to design the circuits they needed. Although complex variable theory lay at the heart of such models, it was not needed for most of the day-to-day work of design engineers.

Although the use of complex numbers appears to have mystified many early electrical and electronics engineers – Steinmetz's Chicago talk is (anecdotally) stated to have been met with much incomprehension, for example – the adoption of the phasor approach was highly significant for future developments in two particular ways. First, as noted by Steinmetz himself in the first quotation above, it allowed quite complicated calculations in the time domain to be replaced by much simpler ones in terms of frequency. Second, it was an important step towards the 'black box' concept – that is, the behaviour of a linear system can be modelled in terms of its input and output without knowing the precise details of its constituent parts. For example, any single-input, single-output network of passive linear components (resistors, capacitors and inductors) can be easily specified by its frequency response derived from those of the individual components. The defining equations for resistors, capacitors and inductors were all subsumed into a generalised, complex version of Ohm's relationship; and even if it would be premature to talk of 'implicit 2-terminal black boxes' at this time, such a representation of components as complex impedances was clearly a great conceptual step.

Furthermore, much greater attention in the manufacture of such devices was subsequently paid to improving the closeness of the approximation of the component behaviour to the ideal mathematical model. A great deal of engineering effort was devoted to ensuring that resistors, capacitors and inductors behaved as linearly as possible over a wide range of operating conditions. Indeed, in this way the mathematical model became a *specification* for the manufacture of devices, rather than an *analysis* of their behaviour. An analogy will perhaps make this clearer. A geometrical progression can be used to model inflation in an economy; this is an *analysis* of more complex behaviour that can be modelled in this way. But if a bank offers compound interest on a deposit, this is a *specification* of how the interest will accrue, based on the same mathematical model.

In terms of the simplest electrical device, the resistor, an *analysis* of various materials shows that Ohm's relationship holds roughly over a given range of conditions. Using Ohm's relationship as a *specification*, however, means that manufacturers have to find ways of avoiding as far as possible deviations from non-linear behaviour. Typically, therefore, materials are chosen – or developed specifically, such as manganin at the end of the 19th century – whose resistivity varies as little as possible with temperature; and the physical construction of the device is designed so as to dissipate heat generated into the environment. Or, when using a wire-wound resistor, care is taken to ensure that inductive effects are negligible under normal operating conditions. The model is influencing – even determining – the technology.

4.2.2 Filter Design and Circuit Theory

In the years following the first publications on the subject by Steinmetz and others, phasor analysis became a basic tool of all electrical, electronics, and communications engineers. During the first two decades of the twentieth century communications engineers made enormous progress in the use of time- and frequency-domain techniques. Heaviside's operational calculus (which converted differential equations into algebraic ones) was put on a much firmer mathematical footing by G. A. Campbell, J. R. Carson, and others. Three closely related approaches were of great significance:

1. The operational calculus itself, which associated mathematical variables in a particularly direct manner with physical quantities, and greatly assisted engineers in the manipulation of such quantities as part of analysis.
2. The mathematical technique of convolution (about which a little more later) used to calculate the input-output behaviour of linear systems in terms of time-varying waveforms rather than frequency responses.
3. The use and dissemination of Fourier analysis to calculate input-output behaviour in the so-called 'frequency domain'.

A driving force for technological development at the time was the need to exploit bandwidth effectively for both carrier telegraphy and the newer telephony. The solution was so-called 'frequency-division multiplexing' (FDM). In FDM, different channels are allocated different parts of the overall bandwidth of a transmission medium. The best-known example is in analogue radio and television broadcasting, where different channels are allocated to different parts of the radio spectrum.

For successful FDM, wave filters with quite stringent pass-band characteristics were needed in order to select the desired channel without excessive distortion. The wave filter was invented independently by Campbell in the USA in 1909 and K. W. Wagner in Germany a few years later. In developing design techniques for such filters, communications engineers started to use mathematics in a radical new way. In particular, circuit diagrams became what we might term a 'meta-language' for the mathematics. Interested readers are referred in particular to the classic 1920s papers by Campbell, Zobel, Foster and Cauer: one of the most accessible introductions to their ideas is still to be found in Guillemin (1935), where full bibliographic references can be found. Complex filters were represented as a series of interconnected sections, then elaborated as a set of equivalent circuit configurations. What was beginning to emerge at this time was the distancing of circuit diagrams in the design phase from their eventual implementation: the manipulation of circuit *component* symbols (in general, complex impedances) became an attractive alternative to the manipulation of *mathematical* symbols. Ultimately this was to lead to the use of prototype circuits in filter design that bear little or no resemblance to the ultimate electronic circuit.

The concept of the linear 'black box' appears to have been made explicit for the first time by Zobel (1924). In this approach, a complex linear electrical network is represented by its input-output behaviour; at this level of abstraction the precise nature of the interconnections of components inside the black box become irrelevant. A modern view of such 'black-boxing' is shown in Fig. 4.2, with particular reference to the relationship between time- and frequency-domain input-output relationships. (This figure is effectively an elaboration and extension of the Appendix to the previous chapter.) This modern view, included here to aid discussion, is significantly different from Zobel's 'general linear transducer' model of 1924, although the general principles are closely related. In the frequency domain, input and output signals are represented by their spectra (Fourier transforms) and the output is obtained by the complex multiplication of the input spectrum by the system frequency response (transfer function). In the time domain, the operation modelling the system behaviour is its impulse response, the convolution of which with the input gives the output. What is important about such models is both the abstraction of the real system into (often comparatively simple, low-order) models, and the equivalence of time and frequency domain models, moving between the two of which gives the designer enormous flexibility and power.

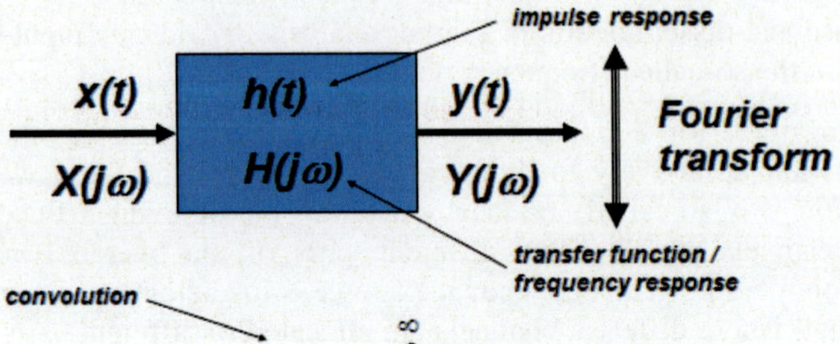

$$y(t) = x(t) * h(t) = \int_{-\infty}^{\infty} x(t).h(t-\tau)d\tau$$

$$Y(j\omega) = X(j\omega) \times H(j\omega)$$

Fig. 4.2 A linear 'black box' defined in frequency and time by input-output relations.

4.2.3 Circuits and Meta-languages

The 'mathematicisation' of filter design – and the corresponding further development of the electrical 'meta-language' – continued apace during the 1920s. As a later commentator put it:

> Foster partitioned the given rational function into a sum of partial fractions that could be identified easily as a series connection of impedances or a parallel connection of admittances [an example of a powerful tool known as duality, a little more about which can be found in Bissell (2004b)]. Wilhelm Cauer expanded the rational function into a continued fraction representing a ladder network. Each method gave two alternative networks, which were called canonical forms because they could always be obtained from a realizable immittance function and because they employed a minimum number of elements. (Zverov 1966, 130)

Cauer's paper, published in 1926, was perhaps the culmination of this stage in the development. He and his contemporaries showed how quite sophisticated mathematics (partial fractions and continued fractions of a complex variable) specifying desired filter behaviour could be translated directly into circuit configurations, in a way that allowed the circuit designer to move at will between the two representations. Increasingly, designers began to think of filter specifications in terms of these so-called 'prototype' circuits that corresponded to the mathematical model, rather than in terms of the mathematical representation itself. It is important to understand that such prototype configurations, even though expressed as networks of inductors and capacitors, are not necessarily the way that the filter itself is finally realised: rather, such prototypes became increasingly a design tool expressed in a language that communications engineers could understand without reference to the mathematical theory they embodied. The fact that this subsequently became common currency in electronics design should not blind us to the novelty of the approach.

This approach went much further in subsequent decades, with considerable blurring between real circuit components and mathematical abstractions. Writing in an editorial in the September 1955 edition of the IEEE[2] *Transactions on Circuits and Systems*, the editor, W. H. Huggins, noted:

> ... modern circuit theory is concerned but little with the circuit as a physical entity and, instead, has become increasingly involved with ... signals ... Thus the usual circuit diagram may be regarded as a pictorial form of a signal flow graph [i.e. an alternate mathematical representation] ... (Cited in Huggins 1977, 667)

As the years went by, electronics engineers introduced a whole set of new, ideal, circuit elements – such as the gyrator, nullator, norator, and supercapacitor – which became just as 'real' to the circuit designers as resistors, capacitors and operational amplifiers. Some of these 'devices' are non-realisable, yet can be exploited in a highly effective way at the design stage. They are 'real' in the sense that they lead to useful ways of talking about the constraints within a network without writing down the corresponding mathematics *per se*. This particular 'pictorial' representation of circuit constraints is very powerful and is as formal as a conventional mathematical description. It is a way of writing the mathematics in terms of connections of components. Moving backwards and forwards between these mathematical and circuit-symbolic domains corresponds to doing very complex mathematics simply by modifying a circuit diagram or using a network transformation. Furthermore, the approach has the advantage of delivering to the designer a very good intuitive idea of what the result of design changes will be.

4.3 The Golden Age and Its Heritage

It was the 1930s and 1940s that saw the development of some of the most powerful 'meta-mathematical tools' of 20th century information engineering (Bennett 1993; Mindell 2002). As discussed in the previous chapter of this volume, Harry Nyquist (1932) analysed the problem of the stability of feedback systems, and showed how a frequency-domain diagram could allow this to be assessed. A simple graph in the complex plane demonstrated not only the absolute stability of a system, but also how close the system was to this stability boundary. At more or less the same time, Hendrik Bode showed how simple, straight-line plots on a log-log grid could give a good approximation to the frequency responses linear systems such as filters and amplifiers.

For example, the simplest first-order linear differential equation is usually written in an engineering context as

$$\tau \, \mathrm{d}y/\mathrm{d}t + y = \mathrm{k}\, x$$

[2] The Institute of Electrical and Electronics Engineers, Inc.

where τ is known as the time constant and k is the gain. If the input to a linear system is a pure sine wave, the output is a sine wave of the same frequency but, in general, with a different amplitude and shifted in phase relative to the input. Bode represented this for a first-order system with unity gain (k = 1) as shown in Fig. 4.3.

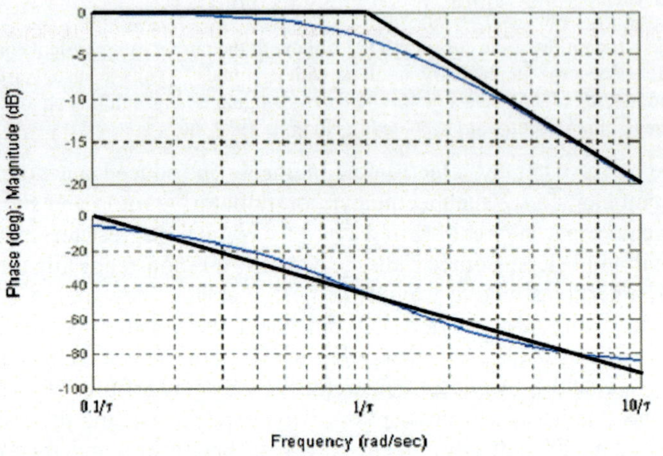

Fig. 4.3 Bode plot of a first order system.

Here the upper part of the figure (magnitude) represents the change in amplitude (measured in logarithmic decibel units) over a range of frequencies, and the lower part represents phase shift over the same frequency range. Bode's key observation was that if the frequency is plotted on a logarithmic scale in units of 1/τ then the precise plot (curved faint line) could be well approximated by the straight lines in bold.

The reader may be forgiven for wondering why this is so important, and a detailed explanation of Bode plot techniques cannot be given here. But as an elementary illustration, consider two such first-order systems, with different time constants, and cascaded (connected one after the other, as might be the case with two electronic subsystems). Then, because of the additive nature of logarithmic representations under multiplication, it becomes a simple matter to derive the overall Bode plot from the constituent parts, as shown for the amplitude ratio in Fig. 4.4. (The annotation 20 or 40 dB/decade gives the slope of the straight lines, where a decade means a change in frequency by a factor of 10.)

Similar techniques can be used for higher order systems, and during the Second World War, Bode's approach was developed further, being applied to generic

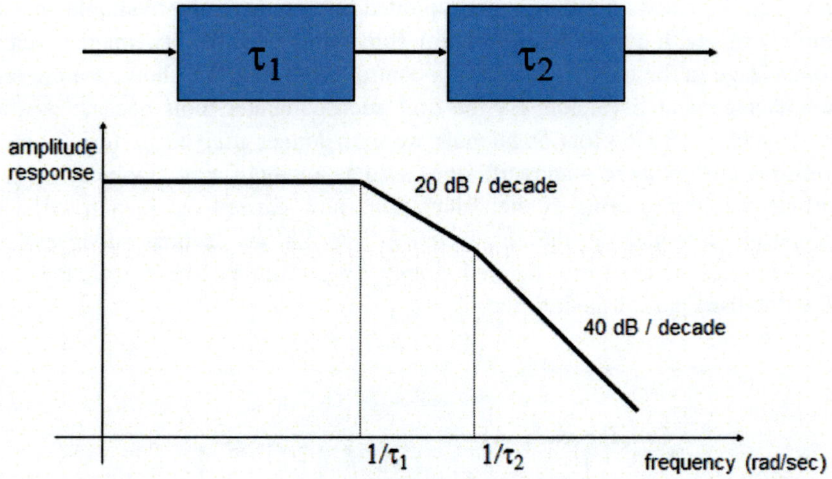

Fig. 4.4 Bode plot of the overall amplitude response of two cascaded first-order systems.

closed-loop control systems such as those used in radar and gun control, as well as electronic circuits. The key point is that, like the Nyquist plot of the previous chapter, this is another way in which engineers have converted quite tricky mathematical manipulations – even solving differential equations – into manipulating simple 'physical2019 objects, in this case drawing short neat lines with a straight edge!

4.3.1 The Nichols Chart

One of the most influential developments was the Nichols chart (1947). By means of this, an open-loop frequency response model, perhaps derived from Bode plots in the way indicated above, or a set of empirical measurements, could be used to assess the corresponding closed-loop behaviour.

For a simple feedback loop such as that in Fig. 4.5 it is relatively straightforward to derive the closed loop frequency response from simple algebraic arguments. However, as the open-loop transfer function $H(\mathbf{j}\omega)$ becomes more complex the computations become more tedious. Nichols came up with a brilliant response to this simple equation, an analysis which essentially originated in an idea of Harold Black in 1927 (Bissell 2004b, 318–320): Nichols generated a chart to derive the closed-loop response given the open-loop one. Fig. 4.6 shows the chart as originally published in 1947. It has fairly widely-spaced grid lines and a restricted area. Fig. 4.7 shows a 1980s version plotted on commercially available specialist graph paper, with quite a fine resolution (this rather poor quality image is scanned from a stage in the manual design of a control loop). Fig. 4.8 shows a more recent computer generated version. By the time such computer tools became available, charts such as Bode plots or Nichols were no longer used to perform computations, rather they were – and still are – used as a magnificent graphical aid to absorbing the implications of the calculations now carried out in a fraction of a second by computer. So the chart itself in Fig. 4.8 has become rather stylised – any numerical results are obtained simply by positioning the cursor, the grid is now a stylised part of the interface.

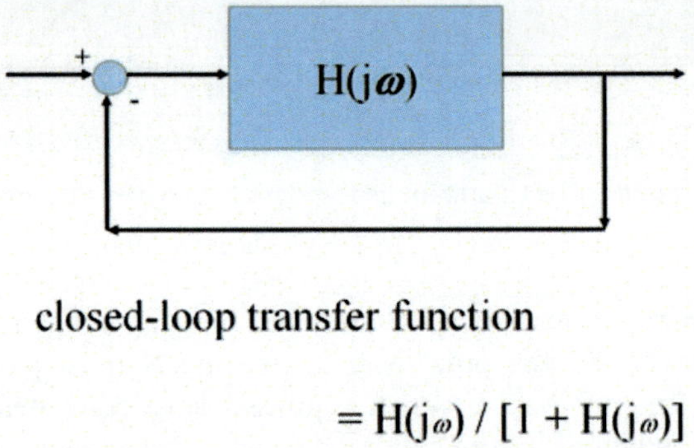

Fig. 4.5 Relationship between the open loop transfer function $H(\mathbf{j}\omega)$ and the closed-loop transfer function (frequency response).

The rectilinear scale is used to plot the open-loop behaviour in terms of the amplitude (in decibels) and phase (in degrees) of a frequency transfer function $H(\mathbf{j}\omega)$, either as an analytic model or a set of frequency response measurements. The closed-loop response $H(\mathbf{j}\omega)/(1+H(\mathbf{j}\omega))$ can then be read immediately – again as amplitude and phase – from the curved lines. Imagine the latter as contours. Then the plotted open-loop-response can be interpreted as a route over a three-dimensional surface, where the height of the contours represents the closed-loop amplitude response of the system. An experienced practitioner can easily get a 'feel' for the closed-loop response just as an experienced map reader can get a 'feel' for the terrain to be walked. Furthermore, changing system gain corresponds simply to raising or lowering the plotted curve – the original post-war users were supplied with a transparent sheet to make this easier; modern computer users simply use the appropriate key, mouse, or slider action.

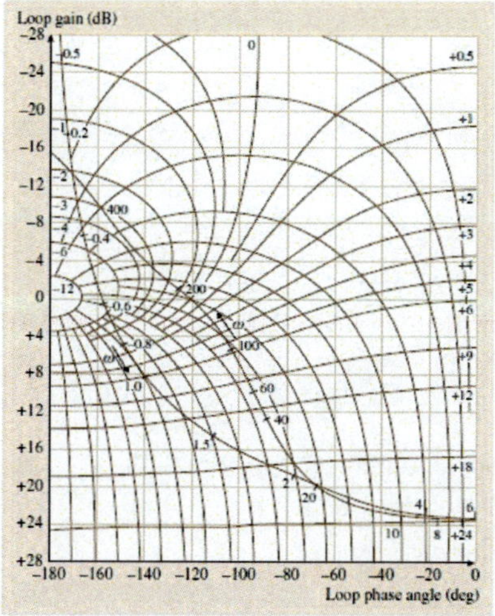

Fig. 4.6 Original form of the Nichols chart. Source: Nichols (1947), reprinted from Fig. 4.8 Springer Handbook of Automation.

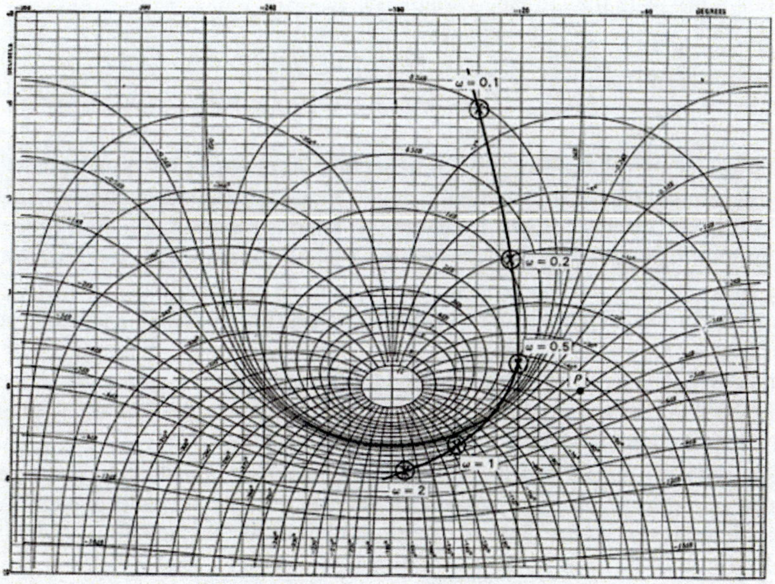

Fig. 4.7 1980s hand-drawn Nichols plot on specialist commercially-available paper.

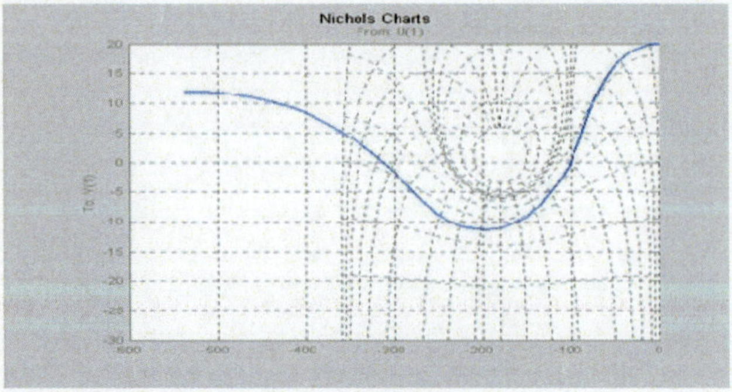

Fig. 4.8 Late 1980s computer-generated Nichols chart.

4.3.2 Poles and Zeros

The late 1940s also saw the appearance of the 'root-locus technique' for the analysis and design of closed-loop systems. A time-invariant, linear system can be modelled by a linear differential equation with constant coefficients. The Laplace transform of this model is a rational function of the Laplace variable (the ratio of two polynomials in the Laplace variable s).

$$H(s)=\frac{N(s)}{D(s)}=K\frac{(s+z_1)(s+z_2)(s+z_3)\ldots(s+z_m)}{(s+p_1)(s+p_2)(s+p_3)\ldots(s+p_n)}$$

The values of s that make this 'transfer function' zero, (that is, when $s = -z_1, -z_{2}, \ldots, -z_{m}$) are called 'zeros', while those that make it infinite (when $s = -p_1, -p_{2}, \ldots, -p_n$) are called 'poles'. Fig. 4.9 shows the simplest example, a first-order linear system with one pole characterised by the transfer function $10/(s + 10)$. Since s is a complex variable we can represent this on an Argand diagram, or *s-plane*, as in the figure. In this diagram the horizontal axis represents the real part of s and the vertical axis represents the imaginary part. The cross on the diagram is at $s = -10$ in this case, at which the value of the denominator of the expression is zero, and hence the value of expression as a whole is infinitely large. The cross, therefore, indicates the position of the pole of the transfer function.

This leads to the powerful root-locus graphical technique. Imagine a surface, corresponding to $|H|$, held up by poles (just one in this case) and pinned down by zeros (there are no finite zeros in this example but the surface drops towards zero at infinity in all directions; hence the function is said to have a 'zero at infinity'). What the root-locus approach did was to show how the poles of a closed-loop (feedback) control system moved around when systems parameters, such as the gain K, were changed. The important point is that the graphical visualisation allowed engineers to 'see' the 3-D surface, and interpret its characteristics in a way that indicated system behaviour.

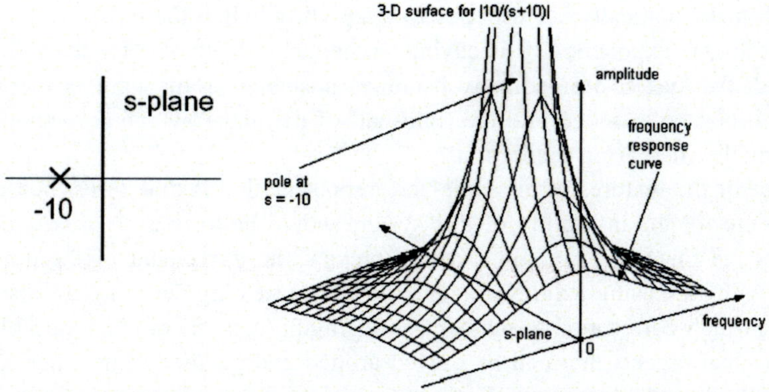

Fig. 4.9 A pole in the *s*-plane at $s = -10$ and the corresponding 3-D surface.

Fig. 4.10 shows a root-locus diagram of a more complicated (fourth-order) pattern of poles and how they move around with varying system gain.

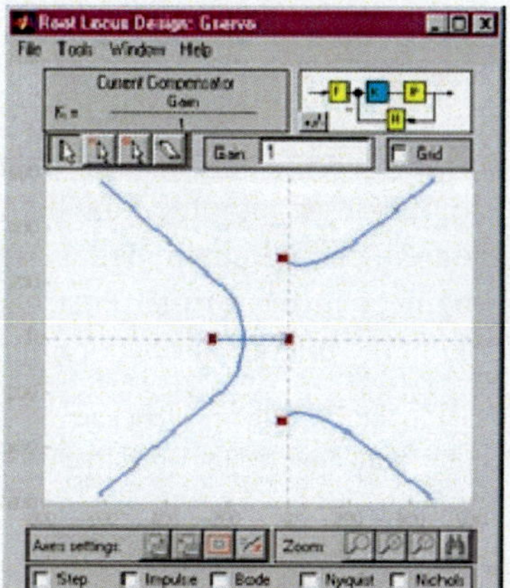

Fig. 4.10 A fourth-order root locus. Image reproduced with permission from Mathworks. © The MathWorks Inc.

The small crosses highlighted show the positions of the poles without the feedback loop connected. In this case there are two, complex-conjugate, poles located very close to the vertical imaginary axis, corresponding to a very lightly-damped oscillation (borderline instability). There is also a pole at the origin (where $s = 0$), which mathematically corresponds to integration, while the left-most pole corresponds to an exponentially decaying component. With the feedback loop connected, the system immediately becomes unstable: as the gain is increased the closed-loop poles move into the right-half of the plane, which corresponds to exponentially increasing oscillations.

One of the features of this pole-zero representation is that zeros 'attract' poles, so a cure for the instability of this system would be to 'place' some zeros to the left of the imaginary axis in order to prevent the movement of the poles to the right with increasing gain. Note, here, how I am treating these highly abstract concepts, which correspond to mathematical singularities in complex variable theory, as physical objects that can be moved around and positioned in order to achieve the desired model behaviour. Later, this model can be converted into an electronic circuit or a computer program and connected to the physical system.

It is easy (once you are familiar with the technique) to convert this type of 2-D model into a 3-D one similar to Fig. 4.9 (but with four poles and a correspondingly more complicated 3-D surface), and thus experience, again, a palpable 'feel' for the system behaviour, in particular the frequency response of the system represented by the vertical cut on the imaginary axis of the Argand diagram.

4.3.3 Models and 'Reality'

In the 1920s and 1930s the modelling approach developed by communications engineers led to considerable unease about quite how the mathematics related to the 'real world'. For some, the frequency-domain models were particularly problematic. John Bray, a senior engineer in the British Post Office (subsequently British Telecom) notes in his book on the history of telecommunications that:

> It seems remarkable now that in the 1920s there were some, including the eminent scientist Sir Ambrose Fleming, who doubted the objective existence of the sidebands of a modulated carrier wave, regarding them as a convenient mathematical fiction ... (Bray 1995, 71–72)

Ambrose Fleming was a pioneer in the development of electric lighting, telegraphy and telephony, and the thermionic valve. He sparked off a vigorous debate in the pages of the journal Nature in 1930, when he called into question the practical significance of the 'sidebands' in the mathematical description of an amplitude-modulated carrier wave. When a sinusoidal carrier is amplitude-modulated by a message signal, the composite spectrum consists of the carrier plus the spec1trum of the modulated signal reflected about it, as shown in Fig. 4.11. Yet Fleming wrote:

> The carrier wave of one single constant frequency suffers a variation in amplitude according to a certain regular or irregular law. There are no multiple wavelengths or wave bands at all. (Fleming 1930, 92)

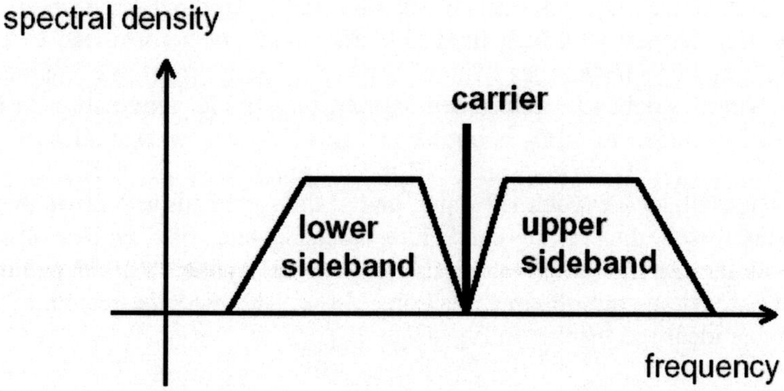

Fig. 4.11 Amplitude modulation of carrier generating two sidebands.

Fleming was certainly well aware of Fourier theory:

> The complex modulation of a single frequency carrier wave might be imitated by the emission of a whole spectrum or multitude of simultaneous carrier waves of frequencies ranging between the limits n + N and n – N, where n is the fundamental carrier frequency and N is the maximum acoustic frequency and 2N is the width of the wave band. This, however, is a purely mathematical analysis, and this band of multiple frequencies does not exist ... (Fleming 1930, 92)

The debate in Nature continued for several months. John Bray returns to this in the conclusion to his book, in a personal reminiscence:

> In 1935 [...] I entered the Open Competition for a post as Assistant Engineer in the Post Office engineering department. This included an interview with the Chief Engineer, Sir Archibald Gill – an interview during which he asked: Do you believe in the objective existence of sidebands? It may seem remarkable that such a question should even have been posed , but at the time there was still a lingering controversy [...] as to whether sidebands were just a convenient mathematical fiction or whether they were 'real'. Luckily I had witnessed [...] a convincing demonstration involving a frequency-swept tuned circuit and a sine-wave modulated carrier, the response being displayed on a primitive electromechanical Dudell oscilloscope and revealed as a triple-peaked curve. So my response to the question was a triumphant Yes, I have seen them! (Bray 1995, 356)

This is an extraordinarily revealing anecdote. To Bray in 1935, the spectrum of the modulated signal displayed on the oscilloscope was just as 'real' as the corresponding time-varying waveform. In fact, both types of display are fairly remote from 'reality: like a spectrum, an electrical voltage or current can only be revealed by an instrument designed to detect it. The fact that the time-domain representation of such variables often seemed (and still seems) so much more natural has more to do with three centuries of natural science and its particular models and conventions than to any ontological distinction.

Communications engineers developed their instruments and practical techniques hand in hand with their mathematical models and other symbolic representations. In a sense, of course, this was not so different from the experience of natural scientists; what *was* different, however, was the way the modelling and instrumentation techniques so developed were soon exploited for *design*. So, for example, the frequency-domain models of modulated waveforms led to a whole range of practical techniques which derived directly from the mathematical models. Single-sideband, suppressed carrier, amplitude modulation originally suggested by Carson in 1915, is one of the most impressive examples of this approach. Carson proposed that, since all the information of the message signal in Fig. 4.11 is contained in each sideband, one of the sidebands as well as the carrier can be suppressed, thus saving bandwidth. The sidebands of a number of message signals can then be combined to use a single channel by means of the technique of frequency-division multiplexing mentioned briefly above. The technique is illustrated in an idealised fashion in Fig. 4.12.

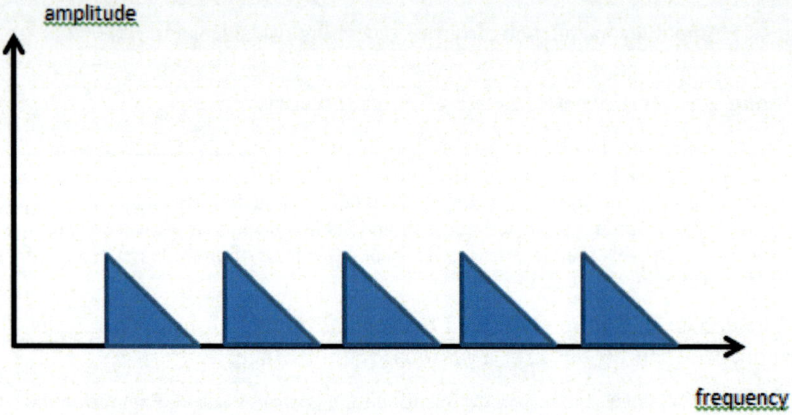

Fig. 4.12 Frequency-division multiplexing of five separate message signals.

Each sideband is translated in frequency by the appropriate amount before transmission, and then recovered by the reverse process on reception. It is particularly interesting to note the language used by communications engineers in describing this process. The sidebands are treated very much as physical objects: they are 'shifted' and 'combined' so as to 'fill' the available bandwidth. They have become something very different from the mathematical spectra (Fourier transforms) of the message signals.

As a final illustration of the heritage of the humble phasor, let us briefly consider the so-called 'signal constellation' model of digital signals. In general, such models are representations of an n-dimensional vector space representing orthogonal digital waveforms and linear combinations of them. Most commonly, however, they are two-dimensional, and represent various combinations of amplitude and phase used to transmit combinations of binary ones and zeros. For example, Fig. 4.13 shows a signal constellation for a modulation scheme known as 32-quadrature-amplitude modulation (QAM). In QAM short segments of sinusoidal waveforms, differing in amplitude and phase, are used to transmit digital signals. In 32-QAM, thirty-two different combinations of amplitude and phase are used, represented by the dots in the diagram, each of which can be considered to define a phasor. Since $32 = 2^5$, each waveform can be used to code a different combination of five zeros and ones. Now, just as a noisy phasor was represented in Fig. 4.1 by an uncertainty in the phasor position, the effect of noise on the QAM waveforms can be indicated by an uncertainty in the precise location of the dots in the constellation. If there is sufficient noise, then one waveform could be misinterpreted as another, leading to errors in the decoded digital signal. The simple signal constellation representation, however, allows certain inferences to be made about the best way to arrange the dots to give maximum separation and thus optimum noise immunity. Again, design engineers 'arrange' the signal constellation in an almost palpable way, even though mathematically we might talk about such arcane concepts as orthogonality and vector spaces.

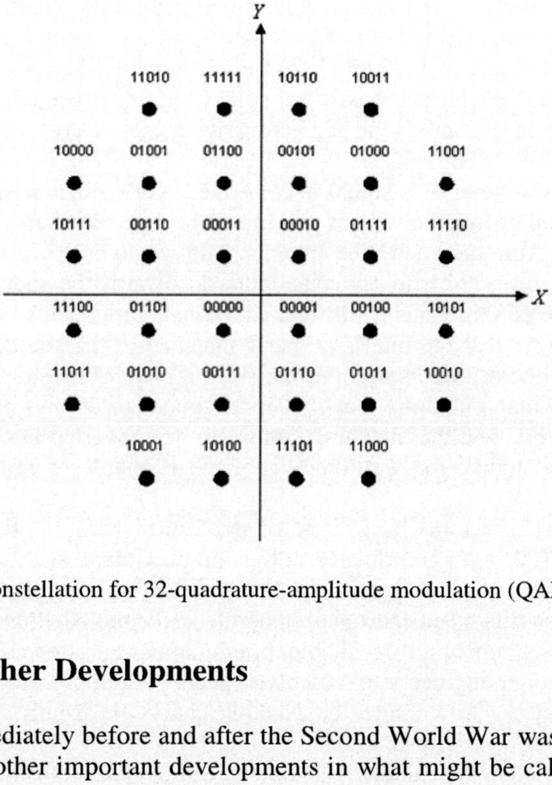

Fig. 4.13 Signal constellation for 32-quadrature-amplitude modulation (QAM).

4.4 Some Other Developments

The period immediately before and after the Second World War was characterized by a number of other important developments in what might be called the epistemology of engineering modelling, directly influenced by communications and control engineering. Four examples will be considered very briefly here: the Smith chart; the rise of general-purpose analogue computing and simulation; the realization of new instruments and devices; and aspects of the modern exploitation of the digital computer.

4.4.1 The Smith Chart

By the 1930s a major design problem for communications engineers was to model the behaviour of transmission lines (various forms of conductors used for high-frequency signal transmission) and to ensure that interconnections were properly 'matched'. Such matching was required in order to reduce or eliminate reflections on the line, or to ensure maximum power transfer between subsystems (depending on the application). Again, the theory was well-understood, but as with a number of application areas in both this and the previous chapter, the modelling tools involve complex mathematics – in both senses of the word 'complex'. For such transmission lines, complex variables are used to model amplitude and phase changes along the line, as well as the input-output characteristics of individual components or sub-systems. In 1939, Philip Smith came up with another stunningly original, powerful – and aesthetically rather beautiful – chart that allowed engineers to derive important results without the need for complex calculation. This is not the place to discuss in detail either the mathematics behind the Smith chart or the precise way in which it was – and still is – used. Essentially, however, it is a way of capturing the behaviour of electromagnetic waves on a transmission line in a two-dimensional diagram.

Suppose the source of a signal is connected, via a transmission line, to an antenna. We want as much power as possible to be transferred from the source to the line, and then from the line to the antenna. And we do not want the electromagnetic wave to be reflected from the antenna back towards the source. In order to ensure that these conditions are fulfilled, additional components can be added to the system to ensure that the line is properly matched to the source and the antenna. This is where the Smith chart comes in, since it enables the complex numbers and mathematical manipulations that model the impedances and other electrical variables of the line and the matching circuits to be converted into curves plotted on the chart – not unlike the way the Nichols chart enables control engineers to do similar things.

Fig. 4.14(a) shows the basic form of the Smith chart. In essence it is a transformation of Cartesian coordinates in the complex plane such that real parts map onto circles and imaginary parts onto the other curved lines. Fig. 4.14(b) is an illustration of how this basic chart was elaborated, like the Nichols chart, into a higher-resolution, commercially available, graph paper that allowed mid-20th century electronics design engineers to 'calculate' accurate component values for particular circumstances. Again like the Nichols chart, the Smith chart is now used primarily as part of a user-interface to computer-aided design programs. The computer does the complex calculations, but the Smith chart helps the engineer in both making design decisions and in interpreting the consequences of, say, introducing a particular 'matching section' between antenna and line. A skilled user can easily interpret typical trajectories of plots on a Smith chart in terms of component values and the general behaviour of the system, and thus derive a solution to a given design task.

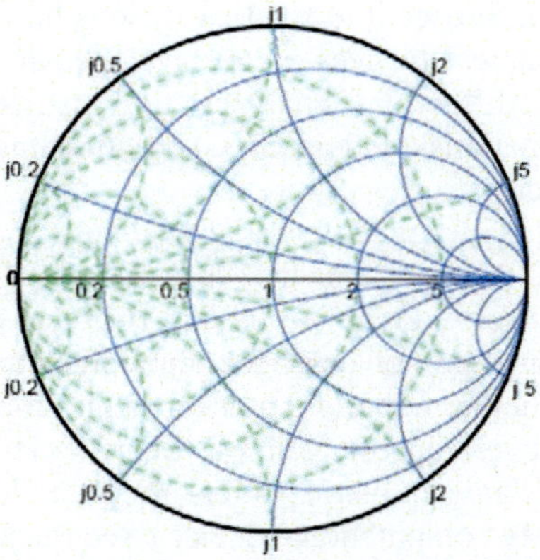

(a)

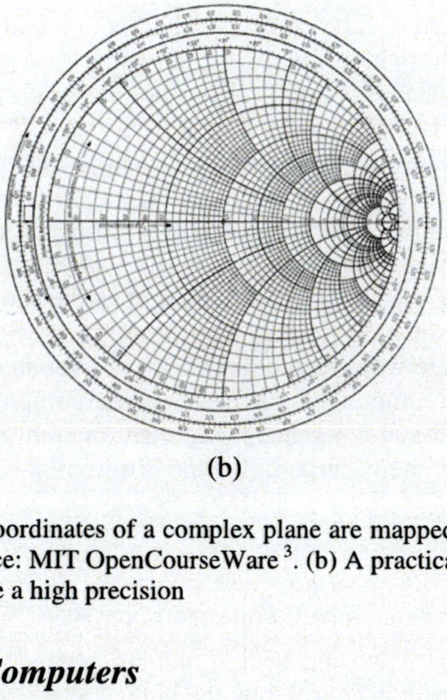

(b)

Fig. 4.14 (a) Cartesian coordinates of a complex plane are mapped on to circles and curves on the Smith chart. Source: MIT OpenCourseWare [3]. (b) A practical design chart capable of generating values to quite a high precision

4.4.2 Analogue Computers

Analogue computing devices – in other words, devices in which physical variables are manipulated directly as analogues of mathematical variables in order to compute solutions or simulate behavior – have a long history. I shall touch only very briefly on this topic here, since a whole chapter of this volume is devoted

[3] Hae-Seung Lee and Michael Perrot, 6.776 High Speed Communication Circuits and Systems, Generalized Reflection Coefficient, Smith Chart, February 2005. (MIT OpenCourseWare: Massachusetts Institute of Technology), Available from http://greenfield.mit.edu/NR/rdonlyres/Electrical-Engineering-and-Computer-Science/6-776Spring-2005/1A878177-DA9E-40A0-9912-F618C1BBE017/0/lec5.pdf). Accessed 30 June, 2011. License: Creative Commons BY-NC-SA.

to analogue simulation. Suffice it to say that various highly sophisticated mechanical and electro-mechanical devices were developed in the 19th and early 20th centuries; the pre-war differential analyzers designed to solve the ordinary and partial differential equations of dynamics and fluid mechanics are perhaps the best-known of the latter.

For a discussion of work on differential analyzers by Vannevar Bush and co-workers see Small (2001) and Bennett (1993). Charles Care's chapter in this volume and his book *Technology for Modelling* (Care 2010) give many additional references and look at types of analogue simulation not considered elsewhere. Norbert Wiener was another important pioneer in this and related fields; for a brief description and further references, again see Bennett (1993). And for a consideration of the interaction between mathematics and technology in the design and development of these early computational devices, see Puchta (1996).

In the late 1940s and 1950s, however, a new generation of general-purpose analogue computers appeared, influenced strongly by the technologies and modelling approaches outlined above. The crucial insight, which had occurred simultaneously in a number of countries during or just after the war, was that the technologies used for high-performance communications and control systems – in essence, the direct descendants of Black's feedback amplifier – could be used to solve a wide range of mathematical problems. In particular, since electronic networks could be realised which implemented quite accurately the mathematical operations of integration and differentiation, suitable interconnections of such networks could be used to solve differential equations directly. And such 'analogue computers' often gave engineers a 'feel' for the problem under consideration very akin to the 'feel' of filter designers for the behaviour of their prototype designs discussed above. An extended meta-language was created in which system behaviour was expressed directly in terms of the components of the analogue computer. As the author of an early textbook on analogue computing remarked: in contrast to the mathematician, who might analyse a system, write down the differential equations, and then solve them,

> the engineer would examine the system and would then build a model of it, which he would call a simulator. He would then obtain solutions of his problem, *perhaps without even having written down the full set of equations* ... [my emphasis] (Wass 1955, 5)

Analogue computers coexisted with digital computers for several decades, even though they were ultimately eclipsed by the latter. In the early days, analogue computers offered many advantages for engineering modelling, particularly in the fields of aerospace and missile control. For example, what they lacked in precision, they gained in speed. A typical application of the immediate post-war 'Project Cyclone', funded by the US Navy, was the simulation of a guided missile in three dimensions. The average runtime for a single solution on the analogue computer facility was approximately one minute, while the check solution by numerical methods on an IBM CPC (Card Programmed Calculator) took 75 hours to run.

The epistemological aspects were just as significant. James Small notes:

> ... the use of electronic analogue computers led not only to the growth of tacit forms of technological knowledge but also to the revision of engineering theory. They enabled engineers to build active models that embodied and operationalised the mathematical

> symbolism of engineering theory. With these models, empirical methods could be used to study the behaviour of a technological system beyond the limits predicted by engineering theory. Electronic analogue computers helped designers bridge the gap between the limits of earlier empirical methods, current analytical techniques and the real-world systems they were constructing. (Small 2001, 274)

4.4.3 New Instruments and Devices

A closely related phenomenon is the way that new, special-purpose devices were produced to implement particularly important mathematical operations. For example, during the Second World War, linear systems theory had made enormous strides, particularly in the handling of random signals and noise (for gun aiming predictors, for example). One particular technique, known as correlation, proved to be extremely fruitful in the theory of signal detection. The correlation function can be written

$$r_{xy}(\tau) = \int_{-\infty}^{+\infty} x(t).y(t+\tau)dt$$

The process of correlation thus involves taking two waveforms $x(t)$ and $y(t)$, and comparing their similarity for varying displacements τ in time against one another. Around 1950, engineers at MIT built an electronic device to carry out this operation and plot the correlation function on a chart recorder. Since then, correlation functions have become just as much a part of the language of communications engineers as spectra, and are implemented in a huge range of modern electronic devices (digital processors proved to be highly amenable to such tasks). Such interaction and synergy between mathematical models, 'meta-representations' and electronic instrumentation was a decisive feature of the development of communications engineering and engineering modelling in general.

In their conclusion to their paper reporting the MIT electronic correlator, the authors remarked:

> The method and technique of detecting a periodic wave in random noise presented here may be regarded as a type of filtering in the time domain [...] From an engineering point of view, many equivalent operations may be more practicable and feasible in the time domain. For instance, a 'zero' bandwidth filter in the frequency domain corresponds to an extremely stable oscillator in the time domain. At the present time it is much easier to build a stable oscillator than to build a network having zero bandwidth. (Lee *et al* 1950, 1171)

Yet again we see the modelling flexibility of moving between time- and frequency-domains, and the way in which the mathematics has become something quite different in the hands of the communication engineers. And, most interestingly of all, perhaps, the mathematical 'black box' entitled a 'correlator' has become a physical black box that delivers ink traces on paper of correlation functions!

4.4.4 Engineering Modelling and the Digital Computer

From the 1950s onwards, the availability of digital computation became increasingly important for the design of engineered systems and their components, as well as for general-purpose engineering computation.

For a long time, the use of digital computers for engineering applications implied a predominantly 'mathematical' approach – in particular the design of robust numerical algorithms and their use for solving differential equations. Over the last decade or two, however, vastly increased computing power, the development of graphical user interfaces, the introduction of symbolic computation, and the ubiquity of the PC have led to a significant change, with a remarkable convergence of design tools, simulation, and implementation. It is particularly ironic, perhaps, that the digital computer, having effectively vanquished its analogue counterpart, has now much in common with the latter in so far as the human-computer interface is concerned. Whereas earlier engineers would physically interconnect operational amplifiers and other devices (implementing in analogue form appropriate mathematical operations so as to simulate a problem), the modern designer is likely to 'drag' symbols representing the various operations into position on screen, and 'interconnect' them as appropriate, before running a numerical or symbolic

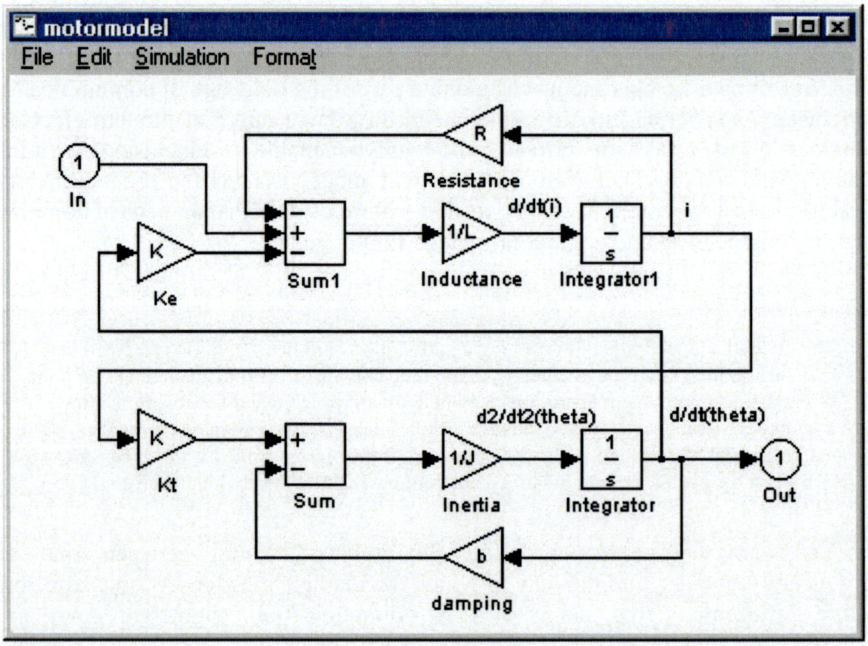

Fig. 4.15 A Matlab© Simulink© model of an electric motor. Note the blocks representing mathematical operations such as integration or addition, plus physical variables such as inductance and resistance. Reproduced by permission from http://www.me.cmu.edu/ctms/controls/ctms/simulink/examples/exsim.htm. Accessed 13 July 2011.

simulation or analysis. What the two approaches have in common is that modern digital computer tools now employ the same 'meta-languages' as the engineer, and the user can again carry out highly sophisticated modelling or simulation activities without employing very much 'mathematics' *per se*.

4.5 Conclusion

In this chapter I have presented a number of what I have called mathematical 'meta-tools' that were developed by electronics and communications engineers over the last century or so. Five themes emerge:

1. Some of the tools are essentially graphical or other pictorial representations that are *isomorphic* in a strict sense with the 'underlying' mathematics – phasors, for example, pole-zero plots in the s-plane, and signal constellations.
2. Some are useful *approximations* to a mathematical analysis that preserve enough of a more theoretical approach to be useful in design while ignoring less relevant complexities – Bode plot linear approximations are an example.
3. Many of the examples convert mathematical theory into much more *concrete* or *reified* representations that can be manipulated in an almost physical manner as part of the design process – almost all of these examples fall into this category.
4. Modern computer tools for engineering design and analysis have combined computing power (to carry out fast and highly accurate mathematical computation or symbolic manipulation) with interfaces that have exploited traditional information engineering charts, techniques and rules of thumb.
5. Finally, I claim that some of the tools developed by these twentieth-century information engineers – in this chapter above all the Nichols and Smith charts – possess a beauty of their own that can also be appreciated as classic examples of graphic design. For those (hopefully including the readers of this book) who can also appreciate at least some of the mathematical and modelling implications and subtleties of these graphics, this aesthetic quality is enormously enhanced.

References

Bennett, S.: History of Control Engineering 1930-1955. Peter Peregrinus, Stevenage (1993)
Bissell, C.C.: Karl Küpfmüller: a German contributor to the early development of linear systems theory. International Journal of Control 44(4), 977–989 (1986)
Bissell, C.C.: Six decades in control: an interview with Winfried Oppelt. IEE Review 38(1), 17–21 (1992)
Bissell, C.C.: Pioneers of Control: an interview with Arnold Tustin. IEE Review 38(6), 223–226 (1992)
Bissell, C.C.: Spreading the word: aspects of the evolution of the language of measurement and control. Measurement and Control 27(5), 149–155 (1994)
Bissell, C.C.: Mathematical 'meta tools' in 20th century information engineering. Hevelius 2, 11–21 (2004a)

Bissell, C.C.: Models and 'black boxes': Mathematics as an enabling technology in the history of communications and control engineering. Revue d'Histoire des Sciences 57(2), 305–338 (2004b)

Bissell, C.C.: Forging a new discipline: Reflections on the wartime infrastructure for research and development in feedback control in the US, UK, Germany and USSR. In: Maas, A., Hooijmaijers, H. (eds.) Scientific Research in World War II. Routledge Studies in Modern History, pp. 202–212. Routledge, UK (2008)

Bissell, C.C.: 'He was the father of us all.' Ernie Guillemin and the teaching of modern network theory. In: History of Telecommunications Conference (HISTELCON 2008), Paris, September 11-12 (2008)

Bissell, C.C.: A history of automatic control. In: Nof, S.Y. (ed.) Springer Handbook of Automation. Springer handbook series, vol. LXXVI, pp. 53–69. Springer, Heidelberg (2009)

Bissell, C.C., Dillon, C.R.: Telling tales: models, stories and meanings. For the Learning of Mathematics 20(3), 3–11 (2000)

Bray, J.: The Communications Miracle. Plenum Press, London (1995)

Care, C.: Technology for modelling: electrical analogies, engineering practice, and the ldevelopment of analogue computing. Springer, London (2010)

Fleming, A.: The waveband theory of wireless transmission. Nature 125(3142), 92–93 (1930)

Guillemin, E.: Communication Networks, vol. II. Wiley, New York (1935)

Huggins, W.H.: The early days – 1952 to 1957. IEEE Transactions on Circuits and Systems, CAS 24(12), 666–667 (1977)

Lee, Y.W., Cheatham Jr., T.P., Wiesner, J.B.: Application of correlation analysis to the detection of periodic signals in noise. Proceedings of the Institute of Radio Engineers 38(10), 1165–1171 (1950)

Mindell, D.A.: Between Human and Machine. In: Feedback, Control, and Computing before Cybernetics. Johns Hopkins University Press, Baltimore (2002)

Puchta, S.: On the role of mathematics and mathematical knowledge in the invention of Vannevar Bush's early analog computers. Annals of the History of Computing 18(4), 49–59 (1996)

Small, J.S.: General-purpose electronic analogue computing 1945-1965. Annals of the History of Computing 15(2), 8–18 (1993)

Small, J.S.: The Analogue Alternative. Routledge, London (2001)

Steinmetz, C.P.: Die Anwendung complexer Grössen in der Elektrotechnik. Elektrotechnische Zeitschrift 42, 597–599; 44, 631–635; 45, 641–643; 46, 653–654 (1893)

Wass, C.A.A.: Introduction to Electronic Analogue Computers. Pergamon Press, London (1955)

Zobel, O.J.: Transmission characteristic of electric wave filters. Bell System Technical Journal 3(4), 567–620 (1924)

Zverov, A.I.: The golden anniversary of electric wave filters. IEEE Spectrum 3(3), 129–131 (1966)

Chapter 5
Early Computational Modelling: Physical Models, Electrical Analogies and Analogue Computers*

Charles Care

Department of Computer Science, University of Warwick, UK

Abstract. The application of computation models and simulations is ubiquitous, their use is key to modern science and engineering. Often described as the material culture of science, models demonstrate how natural it is for humans to encode meaning in an artefact, and then manipulate that artefact in order to derive new knowledge. Today, many familiar models are computational. They run on digital computers, and often demand extensive processing power. In becoming a primary tool for modelling, the widespread use of modern computers has shaped the very meaning of technical terms such as 'model' and 'simulation'. But what did pre-digital modelling and simulation look like? For many, the first technology to support a form of simulation that is today recognisable as computational was electronic analogue computing. Analogue computers were a technology that was in wide use between 1940 and 1970. As early modelling technologies, their history highlights the importance that modelling has always held within the history of computing. As modelling machines, analogue computers and electrical analogies represent a type of computing that was focused on knowledge generation and acquisition rather than information management and retrieval. Analogue technology provides an important window on the history of computing and its use as a modelling technology.

5.1 Introduction

Today, the application of computational models and simulations is ubiquitous. Whether used by an engineer to estimate the structural stress in an aircraft wing, or employed to assist an economist with predicting the impact of a new taxation

* This chapter is based on research undertaken as part of a PhD funded by the Department of Computer Science, University of Warwick (Care, 2008). An early form of this chapter was presented at SHOT 2007, and it develops themes published in Care (2010).

C. Bissell and C. Dillon (Eds.): Ways of Thinking, Ways of Seeing, ACES 1, pp. 95–119.
springerlink.com

policy, models are a key enabler of modern science and engineering. In becoming a primary tool for modelling, the widespread use of modern computers has shaped the very meaning of technical terms such as 'model' and 'simulation'. But was the emergence of simulation practice purely a bi-product of the digital computer age? What did pre-digital modelling and simulation look like?

This chapter explores the history of analogue computing from the perspective of the history of computational modelling. Before the electronic digital computer was invented, the majority of complex computing machines were a type of technology that would now be considered analogue. Analogue computing is quite a broad concept, but these machines generally represented data with a measurable physical quantity rather than abstract symbols (or 'digits'). This chapter looks at two types of analogue computing: indirect and direct analogues. After a look at their history, we will investigate how they were used, and what kind of modelling they facilitated. Even after the invention of the digital computer, electronic analogue computing remained an important computational technology to support modelling. Electronic analogues were in wide use between 1940 and 1970. During this period, technical advances in engineering applications led to the highly-complex emerging technologies of the cold-war period. Analogue computing provided the dynamic modelling technology behind many of these innovations.

Analogue computing represents not just a type of technology, but a broader approach to computing which encouraged users to develop and manipulate models. As a technology, analogue computers have a rich history, and the electronic analogues developed in the 1950s were, in turn, closely related to other pre-computational tools known as analogue models or electrical analogies. These tools evolved from lab-based modelling technologies: their history highlights the importance that modelling has always held within the history of computing. Even after the invention of the electronic digital computer in the 1940s, analogue computers remained important and, for some applications, continued to be used until the 1970s. Much of this later use of analogue computing can be classified as modelling.

5.2 Towards a Use-Centric History of Analogue Computing

When writing histories of technology, it can be difficult to get a handle on how specific technologies were used and understood. To understand the history of computing as a modelling technology, we need to understand the history of both digital and non-digital approaches to computing. The history of computer modelling is as much about the user of the computer, the modeller, as the technology itself.

It is important to understand something of the philosophy and practice of computer modelling. Every discipline has an underlying philosophy: an implicit technical culture binding practitioners and their practices and concepts. A key milestone in the evolution of a new discipline is the point at which practitioners begin to ask questions about why they work in the way they do, what their collective identity is, and what they have to say as a collective. As a window on this philosophy, the personal reflections of members of a user community can provide valuable insights into the technology's history. It is helpful to focus on the writings of some of those who used, and reflected on, analogue computing.

Our analysis begins with an article published in 1972, authored by George A. Philbrick, a character very much entwined with the history of computing.[1] The article was entitled 'The philosophy of models', however, it was not an attempt at abstract philosophy, but more a reflection on how control engineers used electrical and electronic modelling techniques. Philbrick was a practical, intelligent technologist, and a successful businessman. As an 'old timer' of analogue computing and control systems, he used the article to articulate the common-sense modelling culture that was central to the computer's use as a modelling tool. He wrote:

> Computers may... be thought of as general-purpose flexible models or synthesizers, as well as analyzers. The question of names is a controversial issue, involving definitions rather than anything more fundamental, and is most happily resolved by recognition that the equipment under discussion is really a bridge between analysis and synthesis, bringing these two essential modes of study into closer collaboration. (Philbrick 1972, 108)

These words were written right at the end of the era when analogue computing technology was in regular use, and therefore reflect a high degree of conceptual maturity. In particular, we note Philbrick's realization that computing technology has multiple modes of use: namely that computing is both about analysis and synthesis. This is very significant in understanding the relationship between analogue and digital computers. Users typically employ computing technology to both solve explicit problems (akin to analysis) and build more experimental models (more like synthesis). Prior to 1940, analogue computing devices were used for both modes of use. In successive decades, digital computers began to replace them. But the digitalization of equation solving and other forms of analysis happened more quickly than those types of use that were more oriented around modelling, or synthesis. To understand how the modern computer evolved to be used as a modelling machine, it is necessary to understand how modelling was undertaken before digital computing became the dominant form of computing technology.

Before progressing further, it is important to crystallize what we mean by the labels 'analogue computing', and 'modelling technology'. Investigating the connection between analogue technology and computing requires us to unpick what is meant by 'computing', and understand the different forms computing can take. These definitions also need historical context: what we mean today might not have been what people meant in the past. No definition is ever static, and the classification and categorizations of yesterday often shape those of today. Because terminology has evolved over time, the history of analogue computing is a complex web of associations between technology, ideas, and practices.

[1] George Philbrick studied engineering at Harvard and afterwards worked for the Foxboro Corporation on control system simulation. During World War II he was involved in analogue fire control research and afterwards, he set up his own company. This company would become the first commercial manufacturers of operational amplifiers.

Originally, analogue computing was so called because it was a computing technology that employed analogy. This definition originates from classifications made by contemporary users, and the phrase became common around 1940. If we investigate the label 'analogue' we encounter our first mismatch in meaning. Prior to 1940, an 'analogue' referred to something based on analogy (an analogue model relies on analogy or similarity). However, today, the word analogue is generally used to describe non-digital technology (consider analogue electronics or analogue broadcasting). Both these interpretations are valid, and the reason there are two meanings of the word is a direct consequence of how people used, perceived, and described analogue computer technology during the period between 1940 and 1970. During this period, the technologies of electronic analogue computing converged with the technologies of control and signals engineering. These engineering disciplines needed a language to express the difference between discrete or digital signals, and those that ranged over a continuous domain. Through its use by the computing community, the word 'analogue' became a popular antonym, or opposite, of 'digital'. This usage meant that the meaning of the word 'analogue' evolved to become the common label for continuous signals.

Turning to the word 'computing', we find an equally interesting history of meaning and association. As has been discussed at length by historians of computing, the word originally referred to just calculation, or, in other words, numerical computation. Prior to the twentieth century, computers were not machines but the people who undertook the computation. Through the progression of time, computing has become a far broader concept. It now encompasses information and communication technologies: computers are used for document preparation, information storage and retrieval, and communication. Today, the challenges facing computing engineers is not just the automation of mathematical calculation, but the delivery of secure communications, reliable storage, optimized business intelligence and data-warehousing. Symbolic manipulation is still at the heart of what the computing machine does, but through our use of computers, we have attached wider semantics to those symbols. Modern computing essentially encompasses a wide plurality of technology and applications, but historians of analogue computing have not always interpreted analogue technology in terms of these different types of use. It is important to acknowledge that analogue computing was both a calculating and a modelling technology.

One of the key questions in the history of analogue computing is how and when analogue technology was replaced by digital. The answer to this question is complex, and varies depending on the type of application of analogue computing. For numerical processing, the flexibility and accuracy of digital computing replaced analogue technology far more quickly than applications relating to modelling and simulation. Analogue computing had a pre-1940 heyday in calculating and equation solving, but the technology was used for modelling and simulation well beyond 1970.

Models and analogies are clearly at the heart of what analogue computing was used for. But another word existed in the vocabulary of the technology's users.

That word is 'simulation'. Like 'model', simulation is a term that can have many meanings, but it appears to evoke something about the dynamic, interactive nature of a model. A model might be something static, but a simulation is dynamic and often has a temporal dimension.[2] If the key problems of twentieth century science and engineering were managing dynamic systems, the key tool to solve such problems was the simulation of these systems. By their very nature, analogue models were less of an abstraction from the original physical problem, and in many ways set the agenda for simulation practices. When we look at how actual users described analogue computing, we find stories about knowledge construction rather than retrieval, modelling not calculating. For instance, a 1953 report commissioned by the British Aeronautical Research Council described analogue computing as the tool of choice for experimental computing:

> The analogue machine, on the other hand, is more convenient where the problem is itself tentative and experimental; that is, where the choice of later calculations may depend on the results of earlier ones, not in a definite mathematical way, but by the intervention of human intelligence. (Hollingdale and Diprose 1953, 1)

Again we see a division of experiment and processing emerging from the sources. Returning to our earlier quote, this is what Philbrick meant by his use of the words 'analysis' and 'synthesis'. Through the eyes of the 1950s engineer, digital computing appeared to provide the ideal tool for analysis while analogue computing was perfect for synthesis, simulation, or modelling.

5.3 Analogue Computing and the Heritage of Scientific Modelling

Although the phrase 'analogue computer' was first used around 1940, the technologies that received this label had evolved from non-digital calculating machines and 'electrical analogies', or models. While digital computers solved numerical problems, analogue computers were used to solve physical problems. Instead of automating a numerical method, analogue computers modelled a physical scenario that was similar to the problem being studied. That similarity was the 'analogy' from where the technology got its name.

Since the principle characteristic of an analogue computer is that it computes by exploiting an analogy, it is not surprising that the technology evolved, in part, from physical modelling technologies. However, there are many ways of exploiting a physical analogy, and throughout their history, these devices have had a wide range of applications. In general, analogue computers fall into two major categories: firstly, those used to model mathematical systems, and secondly, those

[2] A particularly interesting analysis of time-based analogue computing is given by Morgan and Boumans (2004). In their discussion of the Phillips analogue computer for modelling economic systems, they describe the machine as a four-dimensional model because it represented a dynamic time-based system.

used to model physical systems. Both types of analogue have their roots in nineteenth century science and engineering, and were developed extensively during the first half of the twentieth.

Although the development of analogue computing is an important strand in the evolution of computer technology, the association with computing came second. To understand the history of analogue computing, we see that these devices grew out of specific modelling activities undertaken as part of empirical science and engineering. For instance, one of the most commonly cited examples of early analogue computers is the tide predicting technology developed by Lord Kelvin in the 1870s. Kelvin's tide predicting solution was based around two machines: the harmonic analyzer that processed a tidal curve to extract its harmonic components, and the tide predictor that used tidal harmonic data to extrapolate tide predictions of future tides. The tide predictor employed an innovative combination of pulleys and weights to sum the constituent harmonic components of a tidal curve. As a computing tool, the basic concept of the tide predictor would be adapted and used throughout the world. However, of the two machines, it was the harmonic analyzer which embodied the more complex computation. The analyzer was the first calculating machine to employ a non-trivial mechanism based on mechanical integrators.

Although Kelvin built the machine to mechanize harmonic analysis, the tidal harmonic analyzer could be used to solve more general problems. Kelvin also noted how his invention had the potential to solve differential equations of arbitrary complexity. This vision of a calculating machine to solve systems of differential equations would later be realized in one of the most iconic analogue computers in the history of computing, namely Vannevar Bush's differential analyzer.

5.3.1 Kelvin's Harmonic Analyzer for Tides

On the one hand, Kelvin's harmonic analyzer was, very clearly, a device for solving a mathematical problem. It was a calculating machine, an early computer. However, this machine also functioned as a model of the mathematics behind harmonic analysis. A closer look provides an opportunity to consider just what kind of modelling was being supported.

At the core of the analyzer was a mechanical integrator designed by Kelvin's brother James Thomson. Constructed in brass, the integrator consisted of a disk, cylinder and heavy sphere (see Fig. 5.1). The position of the sphere could be rolled along the cylinder by rotating a geared shaft. The cylinder and disk were also connected to shafts that controlled their rotation. As a calculating mechanism, this combination of components maintained a physical constraint between the three shafts, enforcing a mutual relationship between their individual rotations. Essentially, if the disk were spun with an angular displacement representing a quantity y and the position of the sphere changed to reflect a quantity x, then the cylinder would rotate by an amount corresponding to the integration $\int y\,dx$. The relationship was reversible too, so by controlling the positions of the sphere and cylinder, the mechanism became a mechanical differentiator.

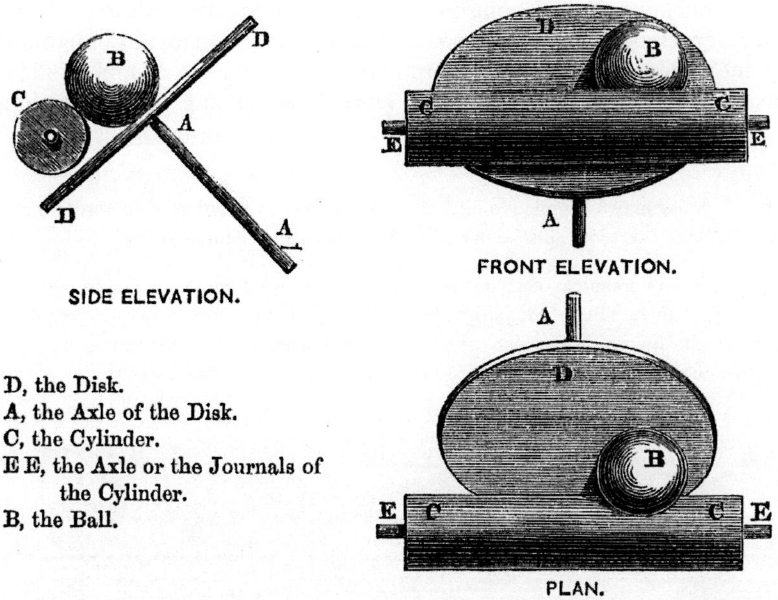

Fig. 5.1 Line drawing of the Thomson disk-sphere-cylinder integrator. Source: Thomson (1876a).

The mechanical integrator developed by James Thomson was certainly innovative, but his invention was in turn inspired by the inventors of a mathematical instrument known as the planimeter. Planimeters were instruments developed for area measurement; many of these mechanisms were invented throughout the nineteenth century.[3] Initially, James Thomson had seen his invention as an improved planimeter. However, Kelvin that realized that the shafts did not have to be driven by human input, but could be mechanically connected together, allowing the result of one integration to be the input into another. Fig. 5.2 shows the harmonic analyzer comprising three interconnected pairs of integrators.[4] Taking this logic further, Kelvin realized that by chaining these devices together, a complex mathematical expression could be modelled in a machine. In a paper presented to the Royal Society, Kelvin wrote:

> Take any number i of my brother's disk-, globe-, and cylinder-integrators, and make an integrating chain of them thus: Connect the cylinder of the first so as to give a motion equal to its own to the fork of the second. Similarly connect the cylinder of the second with the fork of the third, and so on... (Thomson 1876b, 271)

[3] Thomson's integrator was directly inspired by a planimeter designed by Maxwell, which was, in turn inspired by mechanisms exhibited at the Great Exhibition of 1851. See Care (2010, ch. 2).

[4] The harmonic analyzer's integrators were arranged in pairs because they were deriving sine and cosine parameters of the harmonic analysis. Three pairs of integrators enabled three harmonic components to be extracted.

The publication of the Thomson integrator, and this description of how it could be used, represents an important point in the history of analogue computing. It was the point when area calculating instruments became part of an embryonic class of technology that would eventually be labeled 'computing' machines. It is clear from the language of Kelvin's paper that he was well aware of the generic nature of the mechanism. In an additional note he wrote:

> The integrator may be applied to integrate any differential equation of any order. Let there be i simple integrators... thus we have 2i-1 simultaneous equations solved...
>
> Thus we have a complete mechanical integration of the problem of finding the free motions of any number of mutually influencing particles, not restricted by any of the approximate suppositions which the analytical treatment of the lunar and planetary theories requires. (Thomson 1876b, 275)

Fig. 5.2 Harmonic Analyzer for tides designed by Lord Kelvin and comprising of seven Thomson integrators. Source: Thomson (1878).

Although Kelvin never built a fully generic version of a calculating machine based on integrators, his innovation would become iconic in the history of analogue computing. Seventy years later, Kelvin's work on the harmonic analyzer would still be mentioned in the introductory sections of many analogue computing textbooks.[5] The important position he held amongst the later generations of analogue computer users was as much due to his approach to modelling and mechanization as to his technical contribution. Even this analogue computing sales literature from the 1960s situates analogue computing within the heritage of Kelvin's modelling tradition:

> From Leonardo da Vinci to Lord Kelvin, working scale models have been the classic means of examining new designs. Even today they persist in pilot plants and similar trial mechanisms. But their usefulness is limited. Systems grow ever more complex and the laws of Nature do not allow for scaling in the atomic piles, chemical reactors or supersonic aircraft which form the subject of the studies of tomorrow. Today the analogue computer is capable of producing veritable models, truly scaled, of any engineering system which can be represented mathematically, without cost of materials or manufacture, only substituting electrical voltages for the quantities to be measured and networks of resistors and capacitors for the physical structure of the plant or mechanism. EMI (1960)

[5] See, for example: Karplus and Soroka (1959, 152); Fifer (1961, 665); Hartley (1962, 2); Mackay and Fisher (1962, 20 and 171); Welbourne (1965, 4).

For Kelvin, the analyzer mechanism provided a truly dynamic model of the interacting constraints of a differential equation. But what kind of model did this represent? The expression of a computing problem as a mathematical abstraction or formalism sounds very much like modern computing: mathematical formulation followed by a computed solution. However, the role of the mechanization is more subtle. By creating a physical embodiment of the equations, the mechanism provided an animated realization of the abstract mathematics. This was a model in the sense Kelvin was most comfortable: a way of visualizing and experiencing theoretical concepts. As he famously stated in his *Baltimore Lectures*, he saw the ability to build a mechanical model of a theoretical system as an important step towards understanding a system and its theory. Writing about electromagnetism, Kelvin's words provide an interesting insight into how this eminent scientist approached a problem:

> My object is to show how to make a mechanical model which shall fulfill the conditions required in the physical phenomena that we are considering.... At the time when we are considering the phenomenon of elasticity in solids, I want to show a model of that. At another time, when we have vibrations of light to consider, I want to show a model of the action exhibited in that phenomenon. We want to understand the whole about it; we only understand a part. It seems to me that the test of 'Do we or do we not understand a particular subject in physics?' is 'Can we make a mechanical model of it?' I have an immense admiration for Maxwell's mechanical model of electro-magnetic induction. He makes a model that does all the wonderful things that electricity does in inducing currents, etc., and there can be no doubt that a mechanical model of that kind is immensely instructive and is a step towards a definite mechanical theory of electro-magnetism. (Thomson 1884, 111)

> ...I never satisfy myself until I can make a mechanical model of a thing. If I can make a mechanical model, I understand it. As long as I cannot make a mechanical model all the way through I cannot understand; that is why I cannot get the electro-magnetic theory... I want to understand light as well as I can without introducing things that we understand even less of. (Thomson 1884, 206)

Reading this quotation, we hear how the lack of a robust physical model to understand electromagnetism really concerned Kelvin. He wanted to be able to feel the behavior of the equations. Kelvin's words also highlight an important characteristic of nineteenth century physics. In order to understand electricity and electromagnetism, scientists of the day relied on physical models, they understood electrical systems through physical analogies. But intriguingly, as electricity became better understood, people began to reverse those familiar analogies, applying the medium of electricity as a modelling environment to understand physical scenarios. This is where the history of modelling begins to become closely related to the wider history of computing. While analogue computing had exploited analogies in various physical media (including mechanical, hydraulic, and elastic systems), it was through modelling with electrical (and later electronic) state that the modern computer's power would evolve.

5.4 Direct and Indirect Analogues: Mathematical Models vs. Physical Models

Building the harmonic analyzer had shown Kelvin that it would be possible to construct a machine that could model systems of differential equations. Since the harmonic analyzer was based on a chain of Cartesian based integrators, it was not a scale model of the tides but a model of the equations themselves. Integrators modelled the terms in the equations, mechanical linkages representing the relationships between them. Although subtle, this fact would be at the heart of later classifications of analogue technology. Later users would separate analogue computing into two classes of machine. The first class included machines based on mechanical integrators, modelling a set of equations. These were known as *indirect* analogues. The second class were those devices that facilitated a more direct modelling of a physical situation, an interaction more like scale modelling. Machines in this second class were known as *direct* analogues. Both classes of machine facilitated the creation of a model, but because of their use of equations, indirect analogues could also be used for more classic computing problems such as solving systems of differential equations. Under this classification, the harmonic analyzer was an indirect analogue.

Two classic analogue computers are helpful illustrations of the direct/indirect distinction. Both were developed at MIT during the 1930s. They are the differential analyzer and the network analyzer. The differential analyzer was an indirect analogue computer: its physical computing components were connected together to solve differential equations. By comparison, the network analyzer grew out of a need to model power network stability and later evolved into a general tool for modelling other systems with electrical networks. While the differential analyzer solved problems expressed in the language of mathematics, the network analyzer solved problems expressed in the language of electrical networks. However, as discussed below, the differential analyzer was not a purely abstract tool, its physicality in turn encouraged problems to be expressed in the new language of summers, integrators and feedback loops.

During the early twentieth century, the indirect and direct approaches became enrolled into a common discipline of analogue computing. A further convergence occurred when the digital computer was invented. The word 'analogue' was coined to distinguish the technology from digital; both indirect and direct analogues became part of a wider genre of computing technology. Although they eventually converged, the early history of the indirect machines like the differential analyzer was quite distinct from the early history of the direct analogues.

5.4.1 Indirect Analogue Computers and Modelling of Equations

Indirect analogues enabled a user to construct a mathematically oriented model. The resulting system would then become an analogue of a set of equations.

Although this is a general definition, the term 'indirect'was only really applied to those analogue computers based on integrators.[6] While Kelvin had outlined how a generic, integrator-based machine could be designed, he had offered no solution to overcoming the problems of friction and feedback. The central requirement was that integrators should control other integrators, demanding significant torque within the system. The Kelvin harmonic analyzer provided this torque from the sheer bulk of its brass components, limiting the number of integrators that could be chained together. Furthermore, the harmonic analyzer had no way of creating a feedback loop: it was not possible for a single integrator to be connected back onto itself. Over the next half century, the use of integrators evolved significantly. Arthur Pollen used Kelvin integrators in fire control systems, and Hannibal Ford followed a similar path in the United States. Ford's integrator was based on a sphere sandwiched between two disks and held together with springs to provide the required torque.[7]

Vannevar Bush worked on a series of integrator-based calculating machines during the 1920s and 1930s. It was from this work that the differential analyzer evolved. However, the greatest enabler of mechanical analogue computing was probably the torque amplifier. Developed by C. W. Niemann at the Bethlehem Steel Corporation, this invention allowed the integrators to become far more delicate, their output being amplified before driving other components. The torque amplifier also allowed the creation of the all-important feedback loop, enabling the differential analyzer to provide a generic set-up for modelling a system of equations. Although based on equations, the tool embodied the very mathematical relations in mechanics and so offered a new window on differential equations.

The differential analyzer consisted of a number of integrators and plotting tables arranged around a central table. Built into this table were a series of shafts running the full length of the machine, each shaft representing a different variable in the problem under study. Additional shafts, running perpendicular to the others were used to connect the main shafts to a variety of computing components, including summers, integrators, and scalar multipliers (see Table 5.1). Input data was provided by a series of input tables, and outputs were collected by either revolution counters or output plotters. An interesting quality of the differential analyzer was the degree of programmability it offered. The different computing components could be set up in a variety of ways to solve different problems, allowing users to compute with abstractions rather than thinking solely about physical systems.

---

[6] Since the underlying equations form a mathematical model, this type of analogue has the closest mapping to modern interpretations of computing and the modelling of equations gave this class of analogue computer a degree of programmability.

[7] For a more detailed discussion on the history of mechanical integrators and early analogue machines see Bromley (1990), Small (2001), Mindell (2002), and Care (2010).

Table 5.1 Components in the differential analyzer and their mathematical analogues.

Component	Mathematical Representation	Mechanical Implementation
Summer	+	Differential gear: the central shaft rotation is proportional to the rotation of the other two
Integrator	$\int y\, dx$	Wheel and disk integrator: when the input shafts rotates by x and y, the third rotates by $\int y\, dx$.
Scalar multiplication	K * x	Ordinary gear: one shaft rotates proportional to the other. The factor of multiplication was proportional to the size of gear used.

To solve a problem on the differential analyzer, the basic building blocks of different computing components could be used to piece together a system of equations. Each term in the system of equations became realized as one or more components, and the relationships between these terms were codified as physical linkages. The final stage in modelling a problem would be to implement a feedback loop, effectively modelling the equals sign in a differential equation. The resulting linkages established a closed system which would then stabilize on a solution: physical constraints enforced the mathematical relationships. Fig 5.3 shows an example differential analyzer set-up modelling an object in free-fall.

Although primarily designed and used to solve computing problems, the differential analyzer clearly had a constructive model building dimension. Even contemporary writers commented that perhaps the choice of name was incorrect. In 1938, Douglas Hartree suggested the name was 'scarcely appropriate as the machine neither differentiates nor analyses.' Instead, he wrote that the analyzer, 'much more nearly, carries out the inverse of each of these operations.' (Fischer 2003, 87). Another source also commented that a better name might have been 'integrating synthesizer'.[8] Bush clearly saw this simulative function as important. In an address to the American Mathematical Society, he argued that the differential analyzer as an instrument that provided a 'suggestive auxiliary to precise reasoning' (Bush 1936, 649). Many years later he recalled how the physicality of the differential analyzer had enabled one of his technicians to develop an intuitive understanding of differential equations. The mechanical embodiment provided non-mathematicians with a way of feeling, expressing, and exploring the mathematics. Bush wrote:

[8] See Hollingdale and Toothill (1970). They commented that in order to solve a problem, formulae were modelled 'term by term', a method that they considered was 'hardly a process of analysis' (pp 79 – 80).

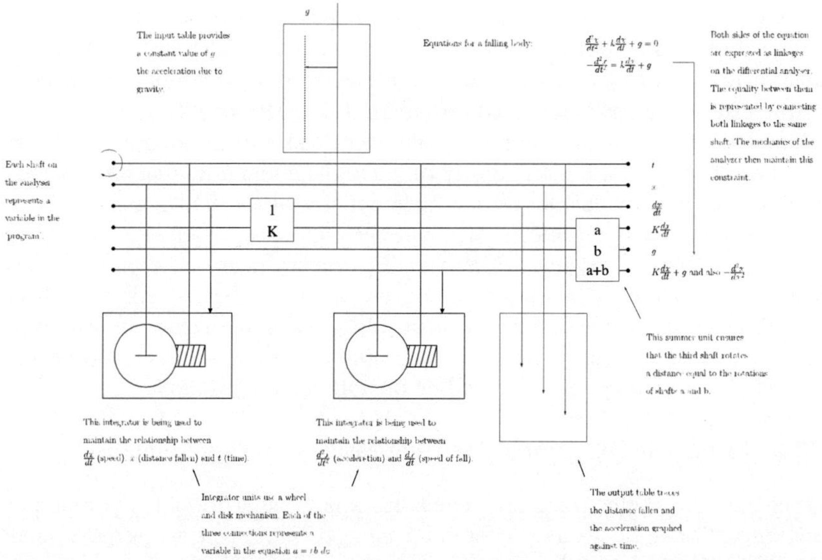

Fig. 5.3 Falling body problem on the differential analyzer based on a problem described in Bush (1931). Source: Care (2010), reprinted with permission.

> I had a mechanic who had in fact been hired as a draftsman and as an inexperienced one at that... I never consciously taught this man any part of the subject of differential equations; but in building that machine, managing it, he learned what differential equations were himself. He got to the point that when some professor was using the machine and got stuck – things went off-scale or something of the sort – he could discuss the problem with the user and very often find out what was wrong. It was very interesting to discuss this subject with him because he had learned the calculus in mechanical … he did not understand it in any formal sense, but he understood the fundamentals; he had it under his skin. (Bush 1970, 262)

5.4.1.1 From Differential Analyzer to Electronic Analogue Computer

The mechanical differential analyzer was certainly one of the cutting-edge computational technologies of its day. During the 1940s it was a vital computational aid in the preparation of ballistics tables for World War II and replicas of the MIT differential analyzer were rebuilt all over the world.[9] The machine also had a role in shaping the future of digital computing; many pioneers of digital computing began their careers using differential analyzers. The American ENIAC, it can be argued, was essentially a digital evolution of the differential analyzer (Birks 2002).

The differential analyzer was an important platform for computing, and during the late 1930s, developing improvements for it was a major research activity in itself. Engineers developed specialized computing components and function

[9] The story of differential analyzer installations and their use can be found in Owens (1986), Croarken (1990), Small (2001), and Mindell (2002).

generators for use in solving specific problems; automatic curve-followers were also developed to automate the input tables. One major attempt to build an automated differential analyzer was the Rockefeller Differential Analyzer which used electronic servo mechanisms to control the linkages (Birks 2002).

Although mechanical computing devices had an important place in the pre-World War II world, the majority of analogue computing machines were electronic. Just as the key enabler for the differential analyzer had been the torque amplifier, the core enabler of electronic analogue computing would be the operational amplifier feedback circuit. Whereas the mechanical differential analyzers represented variables as a shaft rotation, electronic analyzers represented quantities with electrical voltage. Scaling of quantities could be achieved using potentiometers rather than gears. The mechanical integrators and differential gears were replaced by computing circuits based on operational amplifiers.

5.4.1.2 Electronic Differential Analyzers as a Modelling Machine

During the 1950s, electronic analogues became popular tools for computing and modelling. These electronic differential analyzers, or general-purpose analogue computers (GPAC) as they became known, were widely manufactured and installed in many industrial and academic research establishments. Engineering students were taught how to use analogue computers, and many textbooks were published.

One of the advantages of using electronic components was that it meant that an analogue computer could contain far more computing components than the previous mechanical versions. Another major benefit was that the introduction of high speed components facilitated a new type of analogue computing technique known as repetitive operation (or rep-op). Using rep-op, the computer calculated many solutions per second, supporting parameter variation and explorative modelling. This allowed engineers to adjust and tune a model, while seeing the result in real time, perhaps projected onto an oscilloscope screen. Compared to the traditional single-shot techniques of previous analogue computing, this offered a new real-time element to computer based investigation. George Philbrick wrote extensively about this and coined the term 'Lighting Empiricism' to describe this new kind of modelling experience.

> ...models and analog procedures generally can add inspiration to instruction, especially when they can be constructed and operated by the Learner himself in gradual and simple, yet meaningful stages. Talk is fine, symbols on paper are nice, but they are no meaningful substitute for tangible experience with working mechanism.
>
> ...[T]he general principles of Lightning Empiricism... [being] reducing the epochs of trial and error, and learning by experience under conditions where mistakes are not traumatic, and where the results of tentative questions and actions are evident before their purposes have been forgot. (Philbrick 1969, 22–24)

Philbrick's writings offer insight into the analogue computer modelling culture of the 1950s and 1960s. Himself an analogue computer manufacturer, Philbrick clearly had an incentive to promote the technology. However, it is evident that he also had a very clear understanding of what it meant to do engineering modelling using these tools.

> With the practicality of a theoretical proposal established in principle, the experimental stage is entered. In this phase of development there may simply be a filling of detail, or a reduction in the level of abstraction which was maintained on the theoretical plane. This may entail the assemblage and study of more elaborate representing structures: more operational circuits; more OAs [Operational Amplifiers] fetched from the stock room...
>
> Certain criteria for design will be expected to emerge from the experimental phase, if indeed the development project has survived examination so far... By including adjustable imperfections in the circuits of the electronic representation, one finds out rapidly how critical the developing design may be to them... In view of the number of parameters which not infrequently are involved, the attainment of this desirable state can be a formidable task, however praiseworthy the goal, especially without simulative techniques. .. Electronic modelling does not eliminate the need for ingenuity, but it can serve as a valued and uncomplaining partner in the demanding work of design. (Philbrick 1969, 16)

Much of the innovation behind electronic analogue computing came from the control engineering community. Because control systems were also based on these same electronic components, analogue computing provided an ideal environment to model control systems. A technology that had begun as a tool for modelling equations evolved into an important experimental environment for control engineers. This quality of experience was not just articulated by Philbrick. In 1961, Fifer wrote:

> As one learns to interpret the behavior of the computer, one begins to view it as the system itself rather than as some abstract analogue thereof. This resemblance between given and computer systems constitutes the fundamental characteristic which helps to endow the computer with its great value as a design and analysis tool. (Fifer 1961, 2–3)

Although electronic analogues continued to be used until the 1970s, over time they became replaced by software running on digital machines. However, these machines did not represent the totality of analogue computing. In addition to the *indirect* analogues, it is important to consider the tradition of physical modelling that came into analogue computing. The tradition known as *direct* analogue computing includes a number of important analogue techniques that were used during the same period as the better known GPAC machines. They fall into two major types: resistance networks and electrolytic tanks. While analogue computers based on summers and integrators were useful for solving differential equations and problems in control systems, tanks and networks were vital for solving problems involving partial differential equations: problems such as aerodynamics or groundwater flow.

5.4.2 Direct Analogues and Their Evolution from Physical Modelling

The history of direct analogues is a more complex story since they were more special-purpose than their indirect counterparts. However, the sources clearly show that these devices were considered to be an important part of the wider field of analogue computing. For instance, as the author of one comprehensive book on analogue computing writes:

> Two of the most interesting, if not the most accurate, methods for obtaining solutions to the Laplace equation are the soap-film and rubber-sheet analogues. Particularly, the soap-film analogue has been applied to the problem of torsion in uniform bars of non-circular cross section...
>
> The soap-film analogy entails the following procedure. An opening is made in a sheet of metal and the edge of the opening is distorted in such a manner as to make the shape of the opening similar to the cross-sectional area of the bar in torsion... If a soap film is stretched across the opening, the distance of any point... satisfies the Laplace equation. (Fifer 1961, 770–771)

On first reading, this quotation appears far removed from the kind of discourse usually found in a computing textbook. It sounds far more a description of a laboratory experiment. Direct analogues usually exploited some kind of spatial analogy: the shape of a physical model corresponding to shapes in the actual problem domain. For instance, in the soap film example described above, a soap-film frame was constructed to be the same shape as the physical object under study. Similarly, electrolytic tanks involved placing a scale model into an electrical field. Another interesting example of computing using spatial constraints was the Jerie analogue computer for resolving photogrammetric data. The machine was used to resolve common control points in a series of photographs where each image had a degree of overlap and was taken from a slightly different angle and altitude. The Jerie analogue computer could resolve these points by pivoting a series of metal plates inter-connected with elastic bands (Care 2010, 83).

One major application of analogue computers based on physical analogies was the solution of partial differential equations (PDEs). Although standard differential equations could be solved by an indirect analogue such as the differential analyzer, PDEs were generally solved by direct models. One early example was the Laplaciometer. Its inventor, John Vincent Atanasoff, was also a pioneer of digital computing and subsequently invented one of the first electronic digital computers. Atanasoff was a physicist and during the 1930s his research needs inspired him to develop a number of computational aids. Developed in collaboration with Glen Murphy, a fellow physicist at Iowa State College in 1936, the Laplaciometer consisted of a three-dimensional wax model and a four-legged instrument which indicated whether the wax surface satisfied Laplace's equation. In an article on early computing architectures, Burks describes how the device was used:

> To use the Laplaciometer, the computer technician first marked a colored line around the sides of the wax cube such that the height of the line at each point represented the imposed temperature at each point on the edge of the square conducting plate. The technician then carved wax from the cube, working from the top down and working from the sides of the surface towards the center. He continued until the carved surface satisfied Laplace's equation. To test his result at each step of the calculation, the sculptor placed the four-legged instrument on many sample points and observed how close the pointer came to the zero mark. (Burks 2002, 876)

The highly specialized nature of the Laplaciometer leaves it as a bit of an oddity in the history of computing. But this device provides an excellent opportunity to understand the role that modelling played in the way scientists approached computational problems. The example also helps clarify the relationship between indirect and direct analogues. Although mathematical equations still have a role in this instrument, there is no term-by-term modelling of Laplace's equation. Instead, the user of the Laplaciometer attempted to construct a surface that had a mathematical equivalence with the specific problem being studied.[10] In summary, direct analogues were about building up an equivalent model. These tools accomplished this using a wide variety of representations, some of the stranger ones including wax, elastic bands, soap films and rubber sheets (see Care 2010, ch. 2).

Just as indirect analogue technology evolved from mechanical to electronic, the majority of direct analogues were also electrical. As an electrical modelling medium, these direct analogues took two major forms. Analogues were either circuit models: usually networks of resistors and capacitors, or alternatively, they were based on a continuous conductive medium such as conducting paper or an electrolytic tank.[11] Together, these techniques provided tools to build 'electrical analogies' or 'electrical analogues'. Modelling with electrical state required a different set of assumptions; a different set of analogies. It was the shift to electronic representation that would really embed this way of problem-solving into the culture of computing.

Many of the direct analogues evolved from laboratory experiments, and they often looked like laboratory experiments. One such example is the electrolytic tank used to investigate various physical problems. It is an interesting example because it highlights how the knowledge supporting modelling environments evolved over time. As a technology, the electrolytic tank was simply a tool used to explore how electrical charge is distributed in a field. It was an experimental set-up developed during the nineteenth century to investigate and visualize the patterns of electrical fields. The scientists who used the early tanks were not interested in analogue computing, they were interested in understanding electricity. Electricity was something that had previously not been understood, and so it needed investigation. However, as electricity became better understood, people began to identify similarities between physical and electrical systems. During the early decades of the twentieth century, various scientists and engineers identified concrete analogies between physical and electrical problems and started developing electrical models. The electrolytic tank then became a key visualization and investigative tool, not to understand electricity, but rather to visualize electricity in order to investigate something else. Predictable and trustable similarities allowed a tool which had begun as a one-off piece of research apparatus in the nineteenth century to evolve into a modelling technique during the twentieth. In terms of the philosophy of modelling, it is interesting how an instrument of discovery evolved

[10] Philosophically, it might seem that the labels direct and indirect should be reversed. From a mathematical point of view, indirect analogues allowed equations to be modelled directly, and direct analogues required the mathematics to be represented in a physical system (effectively an indirect). However, these labels are not derived from a mathematical perspective, rather from the perspective of physical modelling.

[11] Interestingly electrolytic tanks offered a continuous conductive medium while resistance networks had a necessarily discrete representation. One advantage of electrolytic tanks over conducting paper was that they could be used to model three-dimensional scenarios.

into a visualization tool. It was only through the establishment of concrete theory and trusted analogies that this was possible.[12]

Tanks were first used in a computational sense in aeronautical research to model airflow. In a research note published in 1928, Taylor and Sharman described how they had identified the analogy between flow and electrical patterns. They outlined the details of a tank built at the National Physical Laboratory in Teddington, England. Over the following decades, electrolytic tanks would be used to solve various industrial and scientific problems (Care 2010, 135–149).

One well documented example of an industrial user of an electrolytic tank was Saab, a Swedish engineering firm. Within Saab's aeronautical research laboratories, a tank was built to provide a cheap alternative to wind tunnel experiments. In an article published in 1949, Lennart Stenström, one of Saab's senior research engineers, described the tank they had built and how it was used. To illustrate the use of the tank he cites the example of investigating airflow over the tail portion of an aeroplane. To undertake such a study, the Saab engineers constructed a scale model in Bakelite and investigated how the model distorted an electrical field. Two large electrodes at each end of the tank were connected to an alternating current supply, establishing a uniform charge distribution throughout the tank. When placed in the tank, the Bakelite model, being an electrical insulator, distorted the electrical field. This distortion was analogous to how the actual aeroplane tail would have distorted airflow. In a wind tunnel, air pressures would have been measured using small extraction pipes. On the electrical analogue, the same function was achieved by using small electrodes.

The other major benefit of this kind of apparatus was that the model could be easily modified. For instance, Stenström describes how, after undertaking an initial study on the aeroplane tail, the superstructure was modified by adding modelling clay. Since the clay was also an electrical insulator, the experiment could be immediately replayed to discover the effects of this change. By comparing the electrical field distributions of different shapes, a satisfactory design could be identified quickly and cheaply. Just as Philbrick had used the words 'Lightning Empiricism' to describe incremental design through experimentation, Stenström described the benefits of being able to 'vary the shape of the aeroplane parts progressively'. He wrote:

> The great advantage of the gradient tank as compared with wind tunnels lies in the simplicity of the tank models. At every measuring point on a wind tunnel model intended for measuring the distribution of pressure, a small pressure pipe must be drawn out. The model is expensive and the programme of measurements must be partly known before the model can be constructed. On the other hand, a gradient tank model requires no pressure pipes and is consequently cheap and can be easily subjected to modifications. The

---

[12] In her work on scientific modelling, Mary Hesse (1963) writes about positive and negative analogy. Positive analogies are those aspects that correspond between two situations, and negative ones are those aspects where there is no correspondence, where the two scenarios deviate. Hesse argues that the power of scientific models occurs at the intersection of positive and negative analogies. Models enable the development of theory and help to distinguish the known from the unknown. Computer modelling as engineers use it requires positive and negative analogies to be established and trusted. Only once trust is established can they can be used as a computing tool.

> gradient can be measured at every point on and outside the model without previous preparation. The programme of measurements can be arranged freely and can subsequently be revised without difficulty when measuring results are gradually being available. Thus, it is possible with the help of the gradient tank to vary the shape of the aeroplane parts progressively until a favourable pressure distribution is obtained. (Stenström 1949, 21)

It is clear from this quotation that the electrolytic tank not only reduced the cost of experimental investigation, but also provided a newer, more immediate, mode of interaction. Engineers could modify the model during the experiment and change the location of the probes they were using to make measurements. In retrospect, this use of an electrolytic tank represents the beginning of a simulation and modelling culture based on technology that was relatively compact, real-time, and dynamic.[13] The tank's key strength was that models could be manipulated and results immediately fed back into the design process. Today, modern aeronautical engineers perform simulations via software running on digital computers. But technology only evolves when there is a user demand, or need, for it. Here we can see the 1950s version of that demand: it was the need for engineers to model that led to the creation of analogues like the electrolytic tank, and it was the quality and utility of these analogues that set the requirements which modern software had to deliver.

5.4.2.1 Resistance Networks and Analogue Computing

The electrolytic tanks described above represent computation using analogies based on electrical fields. Another major category of analogue computing was the use of electrical networks as computing environments. Network analogues have their origins in scale modelling power networks. During the nineteenth century, the growth in electrical distribution networks had created a computational demand. It became necessary to analyze stability of electrical distribution grids and model them in a laboratory setting. Thomas Edison, the inventor of the light bulb, had developed a laboratory setup, and a number of model networks were built in the following years. Later networks were not just used for modelling electrical power distribution.[14] In 1908 Edwin Northrup published an article describing the different analogies between circuits and mechanical systems (Northrup 1908). In the following decades work on electrical analogies showed that it was possible to model a variety of problems. For instance, Bush used grids to model stress-strain in civil engineering structures and later built a generic Network Analyzer at MIT. The general framework of a network analyzer allowed a grid of resistors and capacitors to be connected together.

Like the electrolytic tanks, the use of electrical networks generally represents a more direct style of computing rather than the indirect analogues like the differential

---

[13] Only a few feet long, the Saab tank's size is far closer in size to a modern computer workstation than a conventional wind tunnel.

[14] For more information about early developments of network analogues see: Tympas (1996); Tympas (2003); Small (2001); and Mindell (2002).

analyzer.[15] As direct analogues, resistance networks were wired together to represent the structure of the problem being studied. For instance, to model stress and strain in a bridge, a civil engineer would create a network of resistors and capacitors in the same shape as the bridge (Bush 1934). The values of the resistances and capacitances were then chosen to correspond to the physical materials being modelled.

A common application of resistor-capacitor networks was to model hydraulic pressures in groundwater systems, something that was particularly useful in predicting the behavior of subterranean oil reservoirs. The first application to oil reservoirs was by William Bruce in the early 1940s. A physicist working in the research laboratories of Carter Oil, he decided to create an electrical model of an oil field and its aquifer. Bruce's analyzer is an interesting example because it employed both a continuous conductive medium (to model the oil reservoir) and a resistance network to model the hydraulic pressures in the aquifer region around an oil reservoir. Bruce was granted a patent for his invention in 1947 and this approach to modelling reservoirs became widespread throughout the following decades (Bruce 1943/1947; Care 2010, 117–133). Many oil companies developed their own 'reservoir analyzers' derived from the same basic principle.

In fact, modelling groundwater systems or aquifers with electrical networks was one of the applications of analogue computing that persisted the longest. For instance, this technique formed the basis of a major groundwater study undertaken by the British Water Resources Board in 1973. As part of a study investigating the hydrology of the Thames Basin, the board needed to investigate whether artificial recharge of aquifers in chalk rocks in the Thames area could be used to improve water management and supply. Artificially recharging an aquifer involves pumping water underground to increase the groundwater. The purpose of the study by electrical analogue was to investigate how groundwater levels and flow would change as a result of artificial recharge. Their main reason for using an analogue is very typical. The authors wrote that the analogue computer's value was 'in the facility to examine rapidly the broad effects of a wide range of recharge proposals.' (Water Resources Board 1973, iii). Table 5.2 shows the basic electrical analogy employed when modelling groundwater systems.

Table 5.2 Hydraulic/Electrical analogies used in studies of groundwater flow. Based on table in Water Resources Board (1973, 4).

Hydraulic Parameter	Electrical Parameter
Pressure, or head of water	Electrical potential (voltage)
Rate of recharge or abstraction (water in/out)	Electrical current
Transmissivity (resistance to ground water flow)	Electrical Resistance
Storage coefficient (capacity of the aquifer)	Electrical capacitance

[15] An exception to this was the 'transformer analogues' derived from the Mallock machine. See Croarken (1990, 49).

Once again, a closer look at the modelling process involved in modelling an aquifer highlights the spatial nature of a direct analogue model. To construct a model of the aquifer, technicians mounted a map of the region onto a plywood board and drove electrical pins into the board to correspond to the 1km grid printed on the map. This mesh of pins then provided a grid on which to base the resistance network. Using data from observations and geological measurements, the researchers built a model choosing different resistances to model rock 'transmissivity' and connecting those resistances between pins. Known groundwater levels were modelled by applying a voltage to a specific pin. Natural flows out of the system (such as underground rivers) were simulated using electrical circuits that drew current from the grid; transistor circuits being used to model the fact that the flow of a river is proportional to water level. As inputs to the system, the recharge wells were modelled by electrical circuits engineered to generate a constant electrical current. Ordinary wells were modelled in the opposite way – as circuits making a constant drain on the network.

Although the modelling followed an analogue technique, a digital computer was used to process the results. During the run of studies on the model voltages were automatically measured, digitalized, and recorded in a digital format on tape. They were then analyzed by a digital computer (an ICL 1902A) which plotted output charts and contoured maps. It is interesting to understand a little about their technology choices. In their final report, the research board wrote:

> This study was carried out with analogue models, but comparable results could have been obtained with digital computer models of the same system. At the present time, however, it would not be reasonable to model as large an area as the London Basin in such detail, for the computer storage and the running time needed would make costs unacceptably high… As digital computers grow faster and cheaper, and output devices and programs more sophisticated, many of the analogue's traditional advantages of speed and clarity of output are becoming outmoded, and digital computer models may replace analogues in this type of study. (Water Resources Board 1973, 36)

This example provides a window on analogue use right at the end of its era as a common modelling technology. It is fascinating that, even in 1973, a technical report would only state that digital approaches 'may' replace analogue ones. With the benefit of hindsight, it seems obvious that this would have been the case.

One of the digital alternatives to using the resistance network would have been to use finite element analysis. The method of finite elements involves splitting up the domain into many separate sections and then modelling the inter-relationships between these sections. At a high level, this approach feels very digital. However, when compared to the way analogue resistance networks were used we find a high degree of similarity. The analogue resistance network was just like a finite element model: many separate nodes being represented by pins and then interconnected with their neighbors. It is easy to think of analogue as continuous and digital as discrete, but while the resistance network models used continuous physical quantities: voltage, current, charge as the variables of the model, the structure of the model was discrete. And it was this structure that represented the complexity of the computation.

In their report, the researchers of the Water Board had commented that a digital computer would have taken far too long to solve the complexities of this problem. Over time, computers did become more powerful, but by using techniques such as finite elements, the approach taken by engineers to construct and describe the computation was similar to the analysis that underpinned the analogue model. Direct analogue computing was not just about modelling with physical quantities, it was about dealing with the lack of a numerical method; computing with structures that resembled the problem, something that remained important in the equivalent digital approach.

5.5 Conclusion

Models have been employed by scientists for centuries. As a material culture of science, the use of models demonstrates how natural it is for humans to encode meaning in an artefact, and then manipulate that artefact in order to derive new knowledge. Today, many familiar models are computational. They run on digital computers, and often require extensive processing power. However, the role of the computer is far more significant than simply as the engine providing this power. Computer technology has seemingly inspired an emerging culture of modelling and simulation within scientific practice. Computers provide environments for model construction: they are a modelling technology. It is evident that the application of this technology has created new technical disciplines and cultures such as computer simulation and computational science. When computers are considered modelling machines, this reflects a use of technology centered on knowledge acquisition, rather than information processing.

In contrast to digital computing, which is the now dominant form of computing technology, analogue computers manipulated physical state instead of digital abstractions of quantities. They facilitated computation by analogy and were rich environments for modelling. Although many users of analogue computing fully expected digital computing to replace analogue, it was not always very clear what form the digital computer would take. The users of analogue computing wanted to blend the different benefits of analogue and digital computing, and many expected that a hybrid of the two would become common. Hybrid computers were developed, but over time more and more analogue applications were 'digitalized'.

One way that analogue techniques persisted into the digital age was through digital simulation of analogue computers. The first practical software simulation of an analogue computer was developed by R. G. Selfridge, a researcher at the US Naval Ordnance Test Station in 1955. Over the previous years, Selfridge had used a Reeves Electronic Analog Computer. The motivation behind creating his software simulation was to enable problems designed for the analogue computer to run on an IBM 701. Selfridge wrote that 'many of the problems run at present [on an analogue] could be transferred, with advantage, to digital computers.' (Selfridge, 1955).[16] Despite a replacement of the core technology, analogue computing principles remained important, and are found today in tools such as Matlab/Simulink

[16] The history of digital simulations of analogue is discussed in Care (2010, ch. 4).

where it is possible to chain together integrators, summers, and feedback loops in a software application. As noted above, the use of finite element analysis has many correspondences with the way resistance networks were structured.

The histories of both direct and indirect analogue computing show us how practitioners started to rely on more and more modelling abstractions in order to solve real world problems. The evolution of these modelling environments into digital software represented a further such abstraction. This chapter opened with a quote from George Philbrick about analysis and synthesis. When Philbrick wrote these words in the early 1970s, digital computing was finally becoming the common technology to do both. Hence digital computing really did become 'a bridge between analysis and synthesis'.

It is very easy to fall into the trap of perceiving analogue computing as the prior technology of the modern digital computer. However, this is not the case. The development of the two types of technology evolved in parallel. For instance, Charles Babbage's computing machines (designed in the early 19th century) employ principles which we would consider digital. It is true that prior to 1940, analogue computers were the only practical computing technology, but they also persisted for many applications well into the 1970s. Early analogue computers were mechanical but evolved to use electrical and electronic state, and this progression is key to understanding their relationship with computer history. The history of analogue computing is a history of knowledge acquisition and discovery, of theory building and experimentation. It is not just a story about data processing.

In addition, we have seen that the meanings of the words analogue and digital have shifted over time. During the 1930s, the separate themes of analogue computing began to converge. Many publications of that period refer to different 'electrical analogies' of physical systems, but began to refer to such models as 'analogues'. Because they represented state in a completely different way to digital computers, after 1940 it became common to separate the two classes of technology. However, alongside technical differences, different use-patterns started to become part of the classification. For instance, in his *Calculating Instruments and Machines* Douglas Hartree loaded an extra dimension onto the classification. He referred to instruments and machines, a far richer dichotomy, and one that implies a different pattern of usage:

> It is convenient to distinguish two classes of equipment for carrying out numerical calculations... I have found it convenient to distinguish the two classes by the terms 'instruments' and 'machines' respectively... In America a similar distinction is made between 'analogue machines' and 'digital machines'. (Hartree 1949, 1–2)

Instruments are used by someone who wants to discover, reason, and measure; machines offer automation. Hartree goes on to discuss how the 'machines' can calculate to arbitrary accuracy, whereas the instruments are better for getting a rough or approximate answer. Today, many of the analogue technologies are relics of the past, but the problem domains to which they were applied still remain. It is important to see that the need for tools that support speculative and approximate investigation is still an important role for computing to play today. When we look at how users of computing classified their tools, they articulated differences based on both technical characteristics and differences in how the technology was used.

As digital replaced analogue computing, those technical characteristics gradually faded away. However, the different perspectives and approaches of computing are still important today. What makes the history of analogue computing interesting is that we get a window on how those perspectives shifted between different technologies.

References

Bromley, A.G.: Analog computing devices. In: Aspray, W. (ed.) Computing Before Computers, pp. 159–199. Iowa State University Press, Ames (1990)

Bruce, W.A.: Analyzer for subterranean fluid reservoirs. US Patent 2,423,754 (filed September 28, 1943 and granted July 8, 1947)

Burks, A.W.: The invention of the universal electronic computer—how the Electronic Computer Revolution began. Future Generation Computer Systems 18, 871–892 (2002)

Bush, V.: The differential analyzer: a new machine for solving differential equations. Journal of the Franklin Institute 212(4), 447–488 (1931)

Bush, V.: Structural analysis by electric circuit analogies. Journal of the Franklin Institute 217(3), 289–329 (1934)

Bush, V.: Instrumental analysis. Bulletin of the American Mathematical Society 42(10), 649–669 (1936)

Bush, V.: Pieces of the action. Cassell, London (1970)

Care, C.: From analogy-making to modelling: the history of analog computing as a modelling technology. Ph.D. Thesis, University of Warwick, Warwick, UK (2008)

Care, C.: Technology for modelling: electrical analogies, engineering practice, and the development of analogue computing. Springer, London (2010)

Croarken, M.G.: Early scientific computing in Britain. Clarendon, Oxford (1990)

Downing, R.A., Davies, M.C., Pontin, J.M.A., Young, C.P.: Artificial recharge of the London Basin. Hydrological Sciences Journal 17(2), 183–187 (1972)

EMI, Take liberties with time. EMIAC sales brochure. EMI Music Archive: Holdings from EMI Electronics Library. Hayes, Middlesex (1961)

Fifer, S.: Analogue computation, vol. 4. McGraw–Hill, New York (1961)

Fischer, C.F.: Douglas Rayner Hartree: his life in scientific computing. World Scientific Publishing Co. (2003)

Hartley, M.G.: An introduction to electronic analogue computers. Methuen and Co., London (1962)

Hartree, D.R.: Calculating instruments and machines. University of Illinois Press, Urbana (1949); Reprinted, Charles Babbage Institute Reprint Series for the History of Computing. Tomash Publishers, Los Angeles (1984)

Hesse, M.B.: Models and analogies in science. Sheed and Ward, London (1963)

Hollingdale, S.H., Diprose, K.V.: The role of analogue computing in the aircraft industry. Typeset report of the Computation Panel of the ARC. Dated 7th January. National Archives: TNA DSIR 23/21372 (1953)

Hollingdale, S.H., Toothill, G.C.: Electronic computers. Penguin Books, London (1970)

Jerie, H.G.: Block adjustment by means of analogue computers. Photogrammetria 14, 161–176 (1958)

Karplus, W.J., Soroka, W.W.: Analog methods. McGraw-Hill, New York (1959)

MacKay, D.M., Fisher, M.E.: Analogue computing at ultra-high speed. Chapman & Hall, London (1962)

Mindell, D.A.: Between human and machine: feedback, control, and computing before cybernetics. John Hopkins University Press, Baltimore (2002)
Morgan, M.S., Boumans, M.: Secrets hidden by two-dimensionality: the economy as a hydraulic machine. In: de Chadarevian, S., Hopwood, N. (eds.) Models: The Third Dimension of Science, pp. 369–401. Stanford University Press, Stanford (2004)
Northrup, E.F.: Use of analogy in viewing physical phenomena. Journal of the Franklin Institute 166(1), 1–46 (1908)
Owens, L.: Vannevar Bush and the differential analyzer: the text and context of an early computer. Technology and Culture 27(1), 63–95 (1986)
Philbrick, G.A.: A Lightning Empiricist literary supplement 3, preliminary edition. Philbrick/Nexus Research, A Teledyne Company, Debham, MA (1969)
Philbrick, G.A.: The philosophy of models. Instruments and Control Systems 45(5), 108–109 (1972)
Rushton, K.R.: Studies of slotted-wall interference using an electrical analogue: Aeronautical Research Council, Ministry of Aviation, Reports and Memoranda, UK, R & M No. 3452 (1967)
Selfridge, R.G.: Coding a general-purpose digital computer to operate as a differential analyzer. In: ACM, AIEE and IRE Western Joint Computer Conference, pp. 82–84 (1955)
Small, J.S.: The analogue alternative: the electric analogue computer in Britain and the USA 1930-1975. Routledge, London (2001)
Southwell, R.V.: Use of soap films for determining theoretical streamlines round an aerofoil in a wind tunnel. Technical report, Aeronautical Research Council. National Archives: TNA DSIR 23/1710 (1922)
Stenström, L.: The Saab gradient tank. Saab Sonics 12, 18–24 (1949)
Thomson, J.: On an integrating machine having a new kinematic principle. Proceedings of the Royal Society of London 24, 262–265 (1876a)
Thomson, W. (First Baron Kelvin): Mechanical integration of the general linear differential equation of any order with variable coefficients. Proceedings of the Royal Society of London 24, 271–275 (1876b)
Thomson, W. (First Baron Kelvin): Harmonic analyzer. Proceedings of the Royal Society of London 27, 371–373 (1878)
Thomson, W. (First Baron Kelvin): Notes on lectures on molecular dynamics and the wave theory of light. Johns Hopkins University Press, Baltimore (1884)
Tympas, A.: From digital to analog and back: the ideology of intelligent machines in the history of the electrical analyzer, 1870s-1960s. IEEE Annals of the History of Computing 18(4), 42–48 (1996)
Tympas, A.: Perpetually laborious: computing electric power transmission before the electronic computer. International Review of Social History 48(supplement 11), 73–95 (2003)
Water Resources Board, Artificial recharge of the London basin: electrical analogue model studies. The National Archives of the UK (TNA): Public Record Office (PRO) AT 5/36 (1973)
Welbourne, D.: Analogue computing methods. Pergamon Press, London (1965)

Chapter 6
Expanding the Concept of 'Model': The Transfer from Technological to Human Domains within Systems Thinking

Magnus Ramage and Karen Shipp

The Open University, UK

Abstract. 'Systems thinking' is a portmanteau term for a body of theories and techniques that unite around a focus on whole systems and relationships between entities, rather than breaking systems down into their individual components and considering those components in isolation. Various forms of modelling are central within systems thinking, with many of the modelling techniques being developed from work originally carried out in engineering and technology settings, but applied to human-centred application domains, in particular organisations and the environment, but also many others. In this chapter we will discuss four quite different systems modelling approaches that have adapted modelling techniques from engineering to studies of humanity: system dynamics (the work of Jay Forrester and others, applied to organisational, economic and ecological systems); the viable systems model of Stafford Beer (applied to organisational systems); the work of Howard Odum on ecological systems; and the systems diagramming approach of the former Faculty of Technology at the Open University.

6.1 The Understanding of Modelling within Systems Thinking

In this chapter, we will examine a two-stage process of development in the use of modelling: a shift in the domain of application from technological to human domains, and a shift from predominantly mathematical models to more qualitative models. We will trace this development over a period in the 1970s and 1980s, within a range of application areas but especially environmental studies and management.

Our focus in terms of discipline is the area of *systems thinking*. This is a broad term for a set of related techniques and theories that largely arose from the 1940s onwards but unify around the understanding of the world in terms of wholes and relationships rather than splitting entities into their individual parts. Systems thinking has its primary roots in two fields which developed separately at around

C. Bissell and C. Dillon (Eds.): Ways of Thinking, Ways of Seeing, ACES 1, pp. 121–144.
springerlink.com

the same time and gradually came together (along with other fields). The first of these was general systems theory, which derives from the work of Ludwig von Bertalanffy in biology, who focused on processes of organisation, the concept of the open system, and dynamic balance. Secondly, cybernetics derived from wartime work by Norbert Wiener and others on feedback processes in machines and human physiology; in developing as a field, it came to look at feedback and information transfer as part of 'communication and control in the animal and the machine', in Wiener's (1948) memorable phrase. While other strands of work would later feed in to systems thinking, including some of the approaches discussed later in this chapter, general systems theory (GST) and cybernetics formed the main roots of the field. A longer discussion of the history of systems thinking, discussed via its main thinkers, can be found in Ramage and Shipp (2009).

We will discuss here four examples of the way that the two-stage shift of technological to human modelling, and mathematical to qualitative modelling, occurred within systems thinking. First, we will look at system dynamics (in the work of Jay Forrester and his colleagues), which uses large-scale models of feedback derived from servomechanisms to study change processes in organisations, urban settings and the global environment. Second, we will examine the work of Stafford Beer in management cybernetics, which was the first significant attempt to transfer cybernetics into the management domain, doing so by using both mathematical models and analogues of the human brain in studying organisational processes. Third, we will look at models of ecosystems in the work of Howard Odum, who transferred models of electrical circuits into studying the dynamic behaviour of medium-scale ecologies, in the process developing a series of elaborate diagrammatic forms that can be used qualitatively as much as quantitatively. Fourth, we will discuss the little-told story of the way in which the Open University Systems Group developed a mode of teaching systems thinking in a distance-learning context, through using a set of diagram types that had their initial roots in large-scale engineering but were successfully transferred to much smaller-scale human domains and used predominantly as a qualitative tool.

Our focus on these four examples, and the historico-biographical nature of our past work, means that this chapter will take a largely historical stance, discussing the trajectory by which each approach shifted its stance from engineering situations (especially those connected with automation, such as cybernetics and servomechanisms) to human situations. Thus we inevitably focus largely on history rather than the present day. However, this should not be taken to suggest that either the four approaches discussed here, or systems thinking in general, are purely of historical interest. On the contrary, all four approaches have significant present-day user communities and advocates.

Before discussing the individual cases, we will look at the way that modelling has been understood within the field of systems thinking, as this is somewhat distinctive from other uses of modelling. The systems approach to modelling was well described by Donella Meadows, who worked within systems dynamics, largely building environmental models – she was lead author of the celebrated early work *Limits to Growth* (Meadows et al. 1972). She wrote that:

1. Everything we think we know about the world is a model. Every word and every language is a model. All maps and statistics, books and databases, equations and computer programs are models. So are the ways I picture the world in my head – my *mental* models. None of these is or ever will be the *real* world.
2. Our models usually have a strong congruence with the world. That is why we are such a successful species in the biosphere. Especially complex and sophisticated are the mental models we develop from direct, intimate experience of nature, people, and organizations immediately around us.
3. However, and conversely, our models fall far short of representing the world fully. That is why we make mistakes and why we are regularly surprised. In our heads, we can keep track of only a few variables at one time. We often draw illogical conclusions from accurate assumptions, or logical conclusions from inaccurate assumptions. Most of us, for instance, are surprised by the amount of growth an exponential process can generate. Few of us can intuit how to damp oscillations in a complex system.
(Meadows 2008, 86-87)

Perhaps the single defining feature of modelling within systems thinking has been the widespread understanding that models are only a partial representation of reality – they are necessarily incomplete and in the process of abstraction involved in creating models, we both hide and reveal different aspects of the world. This was summed up in a phrase of Alfred Korzybski, made prominent by the influential systems thinker Gregory Bateson, that 'the map is not the territory' (Bateson 1972, 449). This may seem obvious at first sight – on the face of it few educated people would disagree that a model can never be an exact representation of reality. However, in practice, the general public (encouraged by the media) and even the academic community act *as if* models were or should be exact representations. Academics who present models as imperfect or incomplete are popularly regarded as not quite doing their job: witness, for example, the enormous controversy caused by climate scientists' public recognition that aspects of their models were imprecise, taken by some parts of the media to be a denial of climate change.

So if models do not represent reality precisely, what is their purpose? Within systems thinking the answer is that they exist to *sketch* reality and to inform thinking, communication and decision-making, rather than to give an exact answer. In other words, they are a thinking tool. As Pidd (2003, 55) writes, 'models should be regarded as tools for thinking. They provide ways in which people may reflect on what is proposed and decide whether or not certain things should be done and certain risks should be taken'. This is an essentially pragmatic approach, close to the do-whatever-works approach of the engineer, and as such reflects the transition from engineering to understanding human situations that much of the systems thinking discussed here has taken.

An important further aspect of understanding models as thinking tools is that they exist in a particular situation, are constructed from a particular perspective (that of the individual or group who build the model) and with a particular purpose, and are interpreted and used from still another perspective and purpose. If the model was built or used at a different time by different people, then it would look quite different. This is not an argument from strict constructivism, or a suggestion that models have no objective reality – but the details of what is included in a model will vary quite considerably depending on those involved with it. This was well described in a study pack from our own university, which defined a model as 'a simplified representation of some person's or group's view of a situation, constructed to assist in working with that situation in a systemic manner' (Morris and Chapman 1999, 9).

Related to this time- and person-boundedness of models within the understanding of systems thinking, it is useful to observe that some systems modelling techniques are designed in such a way that the models can only be constructed by an expert modeller, while in other cases the models are intended to be constructed in a participatory manner. Some of these models involved (such as the Open University diagrams described later) are specifically intended to be sufficiently simple that a non-expert individual or group can build and use them. This is especially useful in a group context, where a range of perspectives (such as multiple stakeholders in a decision-making environment) can be built into the model. Indeed, in some situations, the *process* of generating the model in this participatory manner – the insights gained from building it – is more important than the product, the actual model itself. Because of this importance of process as opposed to product, and the participatory use of systems models, in many of the techniques used (such as the Viable Systems Model discussed below), models and methods have frequently been intertwined.

A further key characteristic of systems models is that they range from the highly qualitative to the entirely quantitative. Morris and Chapman (1999) distinguish four different forms of systems models: mental models, our taken-for-granted assumptions about the nature of the world, also shaped by language and metaphor; iconic models, such as scale models of buildings, where 'there is usually a strong visual resemblance between the original and the model' (Morris and Chapman 1999, 12); graphical models, two-dimensional representations of a situation such as maps and diagrams; and quantitative/mathematical models, from spreadsheets to computer programs written in specialised modelling languages. As we shall demonstrate in this chapter, the last two types of models are closely connected.

It is worth remarking that not all systems modelling techniques precisely follow the assumptions described above – in particular, some systems modellers would regard their techniques as having a close alignment to reality. Nonetheless, they form an overview of the distinctive nature of systems modelling. We will now explore the four examples of systems modelling approaches in greater detail.

6.2 System Dynamics: From Servomechanisms to Organisations and the Planet

System dynamics is an approach to the modelling of large-scale systems which has had considerable impact in a range of application areas, especially in the United States. Its most notable applications are to organisations and to environmental change, but it has been used to model a wide range of other areas (notably cities and economies). Like cybernetics (with which it shares a number of common features, although they have different roots) it focuses on complex feedback mechanisms within organisationally closed systems. A typical system dynamics model will contain a large number of interconnected variables, measuring the 'stocks' (values) of those variables and the 'flows' (relationships) between the variables.

System dynamics has always been strongly dependent on computer modelling. It was originally developed by Jay Wright Forrester (b. 1918), an engineer at the Massachusetts Institute of Technology (MIT), who was project leader of an influential early digital computer, Whirlwind. Designed for military aircraft flight simulation, Whirlwind was the first computer to run in real-time and to use magnetic core memory (Forrester's invention). Having been involved so early in computers, Forrester felt already by 1956 that 'the pioneering days in digital computers were over' (Forrester 1989, 4), and shifted his focus to business management, studying the way in which his knowledge of technology could better inform management practice.

Forrester's approach, which he first wrote about under the name of 'industrial dynamics' (Forrester 1958), rested on his earlier research on servomechanisms – mechanical devices which control large physical systems through feedback. During the Second World War Forrester worked at MIT on servomechanisms to control gun mounts and radar antennae. His knowledge of feedback behaviour, combined with computer simulation, enabled him to develop a means of modelling industrial processes in a way that captured a great deal of richness and complexity. Feedback control theory enabled him to observe 'the effect of time delays, amplification, and structure on the dynamic behaviour of a system ... [and] that the interaction between system components can be more important than the components themselves' (Forrester 1961, 14).

Forrester's research group, and its output, grew significantly from the 1960s onwards, moving into the dynamic behaviour of urban areas (which later formed the foundation of the successful 1989 computer game *SimCity*). As it shifted into other areas the approach changed its name from 'industrial dynamics' to the broader-focused 'system dynamics', although as George Richardson, a former student of Forrester's, observes, "the word 'system' in the name causes some confusion in identifying the methodological and philosophical ancestry of the field" (Richardson 2000, vii) – it refers to the focus of the approach on analysing the dynamics of "any system" rather than any connection with earlier systems approaches such as general systems theory.

The field which first brought the approach to large-scale public prominence was environmental modelling. Forrester became involved with an international organisational of influential business-people, academics and politicians called the Club of Rome, who were focused on tackling the '*problématique humaine*': a set of interlocking global issues around population, pollution, resource depletion, crime and so on.

The project was carried out by Forrester's research group under the leadership of a married couple, Donella and Dennis Meadows, with a strong passion for global action and awareness of the need for environmental change. The book describing the project, *Limits to Growth* (Meadows et al. 1972), often known as 'the Club of Rome report', sold many millions of copies. It argued that the idea of continual economic growth is unsustainable, in that the global environment is simply unable to cope, and that limits must either be set on that growth by humanity or will be imposed upon us by the capacity of the planet. The book was controversial and contested, and its argument (while accepted more widely today) remains the subject of huge debate. However, a recent re-analysis of the model concluded that "the observed historical data for 1970-2000 most closely matches the simulated results of the [Limits to Growth] 'standard run' scenario for almost all the outputs recorded" (Turner 2008).

The conclusions were already beginning to be argued by a number of people in the 1960s, but the book was distinctive in that it drew those conclusions from large-scale computer modelling. For a book that sold so widely, it is highly technical: it contains many charts and graphs, and several diagrams of feedback loops between variables. The authors are quite open about their method, writing of 'a new method for understanding the dynamic behaviour of complex systems ... the world model described in this book is a System Dynamics model' (Meadows et al. 1972, 31).

While system dynamics grew considerably as an approach over the following decades, spreading far beyond its MIT base to be used internationally, its next point of prominence was to come in the late 1980s. For a number of years, another of Forrester's group, Peter Senge, worked with a group of management consultants to take the approach in a new and softer direction. In his work, published in the widely-read book *The Fifth Discipline* (Senge 1990), system dynamics became the cornerstone of a set of personal and collective disciplines that managers needed to master to build a 'learning organisation', one that is responsive to rapid change in its environment.

What Senge presents in the 1990 book is not pure system dynamics – there are no graphs or computer models of the kind used in Forrester's writing and in *Limits to Growth*. Instead, he describes what he calls 'systems thinking' (a confusing term given its wider base, which Senge has later regretted – Lane (1995)), and which Wolstenholme (1999) refers to as 'qualitative system dynamics'. This version of system dynamics rests on diagrams of feedback loops, often carefully constructed, but no computer models are built from them. Instead, lessons are drawn from the loops, often via the use of 'system archetypes', a set of typical system behaviours with pithy names such as 'Fixes That Fail' and 'Tragedy of the Commons', some of them resting on earlier ideas from other sources (such as the latter, from Hardin (1968)).

The huge success of *The Fifth Discipline* is mirrored by the availability of other works that draw on the lessons of system dynamics in a way that is accessible to the non-mathematical reader, such as *Thinking in Systems* (Meadows 2008), the posthumously published book by Donella Meadows, lead author of *Limits to Growth*. This qualitative approach to system dynamics has been paralleled by an ongoing use of computer models, enabled both by accessible software packages such as iThink and Vensim that easily allow the generation of system dynamic models, and the publication of well-written textbooks with a strong technical background (Sterman 2000; Maani and Cavana 2000).

At the heart of all system dynamics approaches, whether computer-simulated or qualitative, is the causal loop diagram. This shows a set of connected feedback loops made up of chains of variables dependent on each other. Links between variables are tagged with a + or – to indicate whether a rise in the first variable leads to a rise in the second (+) or to a fall in the second (–); the alternative tags 's' (for 'same', i.e. +) and 'o' (for 'opposite', i.e. –) are sometimes used. Loops which show positive feedback are frequently described as 'reinforcing', while negative feedback loops are described as 'balancing'. Fig. 6.1 illustrates a causal loop diagram with a number of feedback loops.

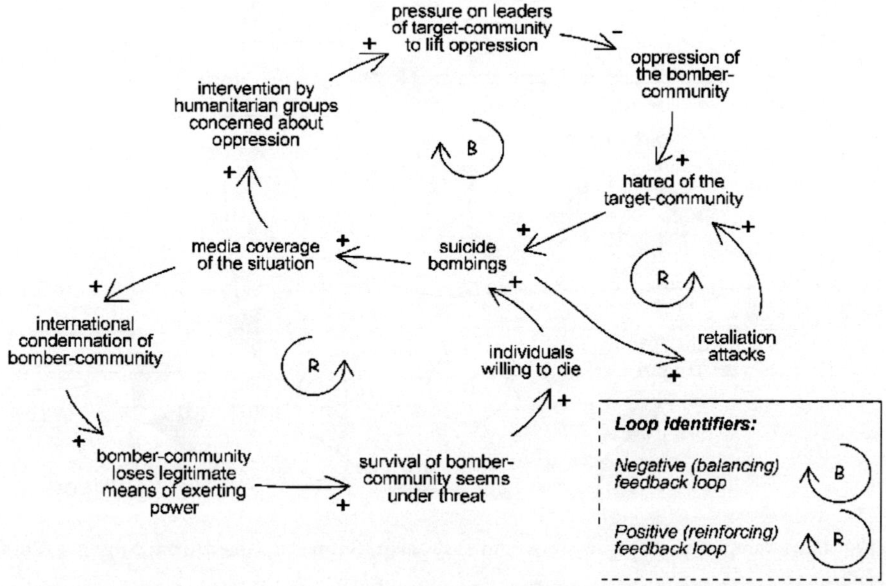

Fig. 6.1 A causal loop diagram showing three possible feedback effects if an oppressed community resorts to suicide attacks.

As shown by Senge's work, causal loop modelling produces rich results that are frequently sufficient to gain deep insights. Practitioners and educators have taught causal loops modelling to a wide range of groups with little mathematical training, to very good effect (e.g. Nguyen et al. 2011). However, many argue that this version of system dynamics needs to go a stage further to build in the power of computer modelling: Forrester himself wrote that 'Some people feel they have learned a lot from the systems thinking phase. But they have gone perhaps only 5 percent of the way into understanding systems. The other 95 percent lies in the system dynamics structuring of models and simulations based on those models' (Forrester 2007, 355).

The computer modelling of causal loop diagrams in system dynamics draws on the stocks and flows model mentioned above. As Sterman (2000) observes, this rests on a hydraulic metaphor – the understanding of variables as flows of water into and out of reservoirs or bathtubs. The diagrams consist of several components: stocks (rectangles, to illustrate containers), inflows (arrows, illustrating pipes, entering the stocks), outflows (arrows leaving the stocks), valves (controlling the flows), sources (clouds, illustrating the origin of the resources in the stocks outside the system boundary) and sinks (clouds to illustrate the destination of outflows outside of the system boundary). These stocks and flows can be readily turned into equations, described in a simple computer modelling language, from which simulations can then be drawn. Fig. 6.2 illustrates a stock-flow model in system dynamics.

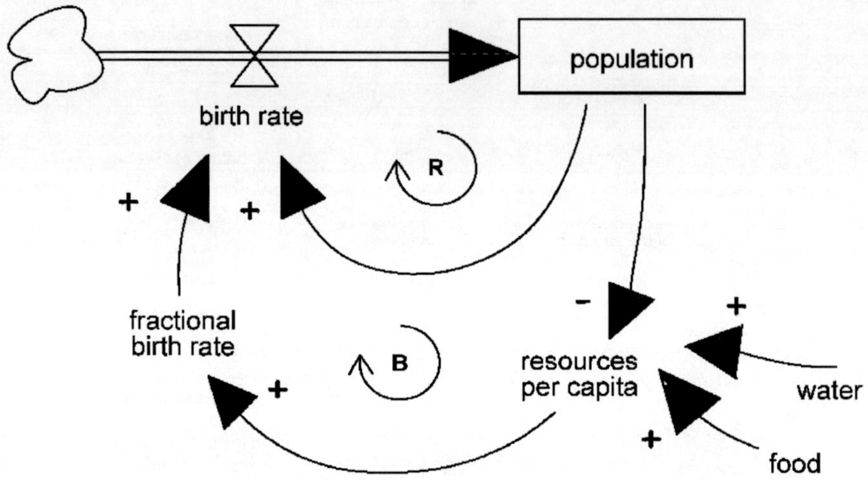

Fig. 6.2 Example of a simple stock and flow map. Source: adapted from Sterman (2000, 205).

System dynamics exemplifies the double transition described in the introduction to this chapter. It had its roots in highly technological disciplines of computation and automation (servomechanisms and early digital computers). These were progressively applied to larger and more complex systems, becoming increasingly human-centred. The shift from quantitative to qualitative can also be readily observed in system dynamics, especially in its use by the practitioner community, although quantitative system dynamics is still highly respected as an academic field and forms the foundation of the qualitative approach.

The successes of system dynamics in a number of areas have led some of its practitioners to assume that it represents the whole field of systems thinking. This is not so, but it does serve as an excellent starting example for the trends discussed in this chapter, which we will go on to see in the remaining cases.

6.3 Viable Systems: From Cybernetics to Management

We turn next to a modelling approach that was again developed in the context of management but, unlike system dynamics, has seldom been used outside of a management context, although it has moved beyond the sphere of business organisations to use in large-scale governmental settings. The Viable System Model (VSM) was developed by Stafford Beer, a larger-than-life British management consultant who developed the concept of 'management cybernetics' as an extension of the principles of cybernetics to management.

The core concept of the Viable System Model is viability, which Beer took to mean the extent to which systems are "capable of independent existence" (Beer 1984, 7). The term is a familiar one in a biological context, for example referring to the capacity of a foetus to exist outside of the womb. In broader contexts it has largely become superseded by the related term 'sustainable', which refers to the capacity of a system to *remain* in existence given changing external pressures put upon it.

The VSM serves as another example of the transfer of technical models and experience to human situations. The technical sources were several: cybernetics, with its mixture of machine and animal (the primary source for the work); operational research, with its mathematical models applied to practical situations; and the somewhat mechanistic form of neurophysiology that was then dominant (which in turn had connections to cybernetics through the work of authors such as Warren McCulloch, a neurophysiologist who chaired the Macy conferences on cybernetics, and whom Beer regarded as a mentor). Beer also took inspiration from prototype and theoretical machines built by himself and others, designed to test various questions, such as the nature of learning in the human mind (Pickering 2004). The application to the human domains of business management and government is very clear, and Beer's primary goal in his work; but the other part of the transformation outlined in this paper was also somewhat present, as the model took its final form in a set of diagrams.

Rather more strongly than some of the models we discuss in this chapter, the VSM was very largely developed by a single individual, so it is worth briefly discussing his background. Stafford Beer (1926–2002) first entered management as a captain in the Gurkha Rifles in India. Upon leaving the army in 1949, he worked for a division of the UK's then-largest steel company, United Steel, where he set up the company's operational research and cybernetics group. He left that company in 1960 and subsequently worked largely as a management consultant, first with a company he established, then after four years working for a publishing company, as an independent consultant. From 1970 to 1973 he carried out his most influential piece of consultancy, working in Chile with the democratic-Marxist government of Salvador Allende to restructure that country's economy on cybernetic lines. Despite the end of that government in a brutal military coup (following which Beer, who had been wealthy and successful, renounced his material possessions and moved to a cottage in rural Wales), he later worked with several other commercial organisations and governments, especially in Latin America. In addition to his practical work, Beer was deeply involved in the theoretical development of management cybernetics. He published several books, was president of international learned societies in systems, cybernetics and operational research, and was a visiting professor at several universities. More information on Beer's life and work can be found in Ramage and Shipp (2009) and Rosenhead (2006).

Beer's conception of the viable system rested on the work of Ross Ashby, a psychiatrist who carried out important early work in cybernetics and wrote the first textbook in the field (Ashby 1956). Central to Ashby's work, and to Beer's development of it, was the concept of *variety*, defined as the number of possible states that all the variables in a system combined can take, and thus forms a measure of the complexity of a system (Beer 1974). From this concept, Ashby developed a theory of the level of complexity necessary to regulate a system, the Law of Requisite Variety. This law states that the regulatory part of a system must contain as much variety as the part of the system being regulated. This does not mean that a financial regulatory agency (for example) must contain the same level of staff and possible states as the financial institutions which it regulates, but it must contain sufficient complexity to be able to control all the possible states and transactions of those institutions. Beer (1984, 11) wrote that 'it has always seemed to me that Ashby's Law stands to management science as Newton's Laws stand to physics; it is central to a coherent account of complexity control'.

Ashby derived his ideas from two key sources: his psychiatric practice (mental processes form a key part of his work) and his development of analogue computers, most notably the Homeostat. This was intended in a mechanical device to mimic the biological processes of homeostasis, and ultimately to model the learning processes of the brain with its capacity to learn from the environment, but to remain in a steady state. It formed the basis for Ashby's second major book, *Design for a Brain* (Ashby 1960), and in the insights it gave into a system's capacity to adapt to change, it led clearly into Beer's conception of a viable system.

By his own account, Beer went through several attempts to describe the model of a viable system in a way that was both theoretically rigorous and practically useful. In the first place he expressed it in terms of mathematical set theory (Beer 1959); in the second, in terms of a neurophysiological model that Beer always insisted was not 'merely' analogous but really drew on a commonality in the underlying structures within both organisations and brains (Beer 1981); and lastly in terms of a set of highly-complex, rigorously defined, diagrams (Beer 1985). It is this final form – the VSM as a set of diagrams, looking a little like circuit diagrams (see Fig. 6.3) – that has endured as the mostly widely-understood form of the VSM. It is worth observing that in Beer's view, regardless of the changing form of the VSM, it was always conceived 'in terms of sets of interlocking Ashbean homeostats. An industrial operation, for example, would be depicted as homeostatically balanced with its own management on one side, and with its market on the other. But both these loops would be subject to the Law of Requisite Variety.' (Beer 1984, 11).

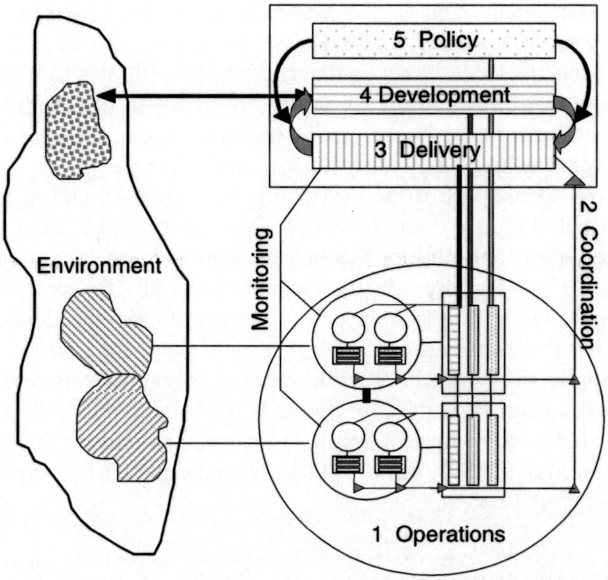

Fig. 6.3 Example of a Viable Systems Model diagram. Source: Hoverstadt (2010, 89).

As a model of a viable system, the VSM consists of five subsystems (referred to as Systems One through Five), which in turn fulfil the functions of implementation, coordination, control, planning and policy-making. Beer defines a clear set of communications and monitoring channels between these sub-systems, and shows the ways in which they must be in a homeostatic relationship to maintain the viability of the overall system. There are two important further features to the model: first, it is recursive, in that 'any viable system contains, and is contained in, a viable system' (Beer 1984, 22). Second, it is not hierarchical: while the five subsystems may superficially look like layers of an organisational hierarchy, they are to be considered as roles, with one person or group able to hold more than one role. Nor is one role to be considered 'higher' than the other. In a story Beer (1981, 258) liked to repeat and incorporated into his later thinking, he presented the VSM and its five subsystems to Salvador Allende, who on seeing System Five (the policy-making role) said 'Ah! *El pueblo*! [The people]' – not the president or the chief executive.

The VSM has had many critics. It is complex, written up in a series of at times quite hard to follow books, rather focused on the structure of organisations rather than any other aspect of their operation, and risks being somewhat rigid – Beer's ultimate definition of viability is that a system matches the structure of the VSM. Also, despite Beer's own advocacy of decentralisation, adaptability and freedom – and his close working with the democratic Marxists in Chile – there is a risk of the VSM lending weight to those within existing power structures who wish to strengthen their power through controlling others. As Jackson (1989, 435) summarises these criticisms, 'the imposition of a particular design may become fixed and prevent necessary adaptation … the VSM can easily be turned into an autocratic control device serving powerful interests'.

A consequence of these issues is that the VSM, while it was used in many different organisations by Beer himself and has been taken forward by a small group of his close colleagues, has never quite had the impact of other major systems models and approaches. It is widely known and respected in the literature, and at least in the UK forms part of the curriculum in systems thinking in a number of universities (including our own university – see Hoverstadt (2010)), but it has had less practical impact than it might have done.

However, for the purposes of this chapter, the VSM forms a very interesting artefact as a further example of a model, firmly within the domain of systems thinking, which drew its inspiration from several technical sources, but shifted its domain of application towards human systems (organisations and government), and its primary means of expression towards the largely qualitative (the diagrams exemplified at Fig. 6.3). It drew on different sources and produced different results from system dynamics, but exhibited the same trajectory, which we will also see in the following two examples.

6.4 Energy Systems Language: From Electrical Circuits to Ecosystems

One of the classic scientific approaches is to interpret the world through a single key factor. This could be regarded as reductionist or over-simplistic, but in some forms of modelling (where abstraction away from detail and towards key factors is crucial) it can prove to be highly productive. This was so in the case of our next example, the Energy Systems Language (ESL) of Howard Odum, who carried out the bulk of his work in the field of ecology and took ecosystems as his primary area of interest, but argued that his approach was strongly generalizable. As we will show throughout this section, Odum's work makes for an interesting comparison with the system dynamics and viable systems approaches already discussed.

For Odum, the single key factor through which he viewed the world was *energy*. He wrote that 'the energy language is a way of representing systems generally because all phenomena are accompanied by energy transformations' (Odum 1983, 5). Odum constructed a sophisticated series of diagrams – the energy systems language – to model energy flows within systems. It appears from Odum's diagrams and writing that he conceived of energy in a predominantly physical way, rather than as a form of metaphor. It might seem that this would limit his approach to only being applied to the physical sciences, such as the ecosystems with which he was principally concerned, but in fact he did attempt to generalise his work.

As with Stafford Beer and the VSM, the ESL was very much the individual product of Howard Odum's work over many decades, and so his life and its influences on the language is worth examining. Howard Thomas Odum (1924–2002) was the son of a prominent sociologist, Howard Washington Odum, and to distinguish the two he was always known as Tom or HT. The elder Odum encouraged his two sons, HT and his brother Eugene, to enter science, and both became prominent ecologists – Eugene Odum was author of one of the key textbooks in the field (Odum 1971), which drew on and popularised HT's energy systems diagrams. As a child, HT Odum was highly influenced by marine zoology, but also by a 1913 book called *The Boy Electrician* (as discussed by Taylor 1988), which would have an important influence on the development of the ESL.

HT Odum spent most of his life as an academic, largely at universities in the southern states of the USA (North Carolina, Texas and Florida), with the largest part of his career at the University of Florida. He carried out detailed fieldwork in a range of ecological settings, including freshwater springs, coral reefs, large ocean bays and tropical rain forests. Some of his more notable field sites included a river ecosystem in Silver Springs, Florida (the site of one of his earliest uses of energy systems diagrams, in Odum (1957)); the US atomic testing site at Enewetak Atoll, in the Marshall Islands; the delicate south Florida wetlands; and the entire Gulf of Mexico. He was a much-loved teacher of generations of American ecologists (Brown et al. 2004) and intellectual leader – he is widely credited as being one of the founders of systems ecology and ecological engineering. He was chair of the International Society for Systems Sciences and (with his brother Eugene) received the Crafoord Prize, the premier international award for ecology.

Energy was at the heart of his theoretical work, as well as his systems models. He developed the maximum power principle, derived from the work in thermodynamics of Alfred Lotka, which suggested 'systems prevail that develop designs that maximize the flow of useful energy' (Odum 1983, 6). He also created the concept of 'emergy' (embodied energy), based on an argument that to determine the true cost of a product or service, we must map the entire energy required at each stage of its production and not just its final energy costs – a familiar idea today in a world of carbon footprints and lifecycle CO_2 costs, but an unusual one in the early 1970s when it first arose.

Detailed discussions of the development of the Energy Systems Language – originally called the energy *circuit* language by Odum, immediately demonstrating the influence of electrical circuit diagrams, and occasionally later known as Energese – can be found in Brown (2004) and Taylor (2005). In its earliest form, in discussing the Silver Springs work, it consisted of a diagram of a river system, showing flows of energy in and out of the ecosystem. As Taylor (2005, 66) observes, 'measurement was central to Odum's ecology ... by collecting data for an entire system and summarising them in flow diagrams, the systems ecologist could act as if the diagrams represented the system's dynamic relations'.

With his boyhood electrical interests, it was a natural next step to convert these energy flow diagrams into an existing, highly developed, diagrammatic representation of energy – an electrical circuit diagram. By modelling ecosystemic energy flows as equivalent to electrical circuits, Odum was able to simulate the behaviour of those flows, first in an analogue computer and subsequently as digital computer programs.

Although Odum continued to use both flow diagrams and electrical circuit diagrams to model energy flows for some years, from the mid-1960s onwards he developed a set of more stylised and abstracted diagrams for his modelling. The symbols in these diagrams represented such factors as energy sources, energy storage, heat sinks, various forms of energy transformations, and so on. By the mid-1970s, they were well developed and sufficiently standardised that Odum produced sets of green plastic templates to enable users of the diagrams to draw the symbols easily. Around the same time, Odum recognised that there was considerable overlap between his symbols and those of system dynamics – both drew on a hydraulic metaphor of flows and tanks, and both relied heavily on feedback. He published two papers comparing the approaches, and made some use of Forrester's Dynamo compiler in his simulations, but the two approaches have always remained separate.

Odum's ESL diagrams became increasingly sophisticated both in form and in content. As Brown (2004, 91) observes, by the early 1970s, 'it was nothing for diagrams to have dozens of compartments and processes, with lines traversing the page. Often publications reverted to fold out pages in order to accommodate gigantic diagrams summarizing everything believed important about one system or another'. In terms of the form, ESL reached its high point in the book *Systems Ecology* (Odum 1983), which covers the form of the language – generally referred to as energy circuit language in the book – in great detail, as well as discussing many other systems modelling languages. In its mature form (see Fig. 6.4), ESL became a very powerful way of modelling the dynamic behaviour of ecosystems, via an understanding of their energy flows. As was the case with the final form of the Viable Systems Model, Odum viewed his diagrams as mathematically rigorous ('picture mathematics', as described by Brown (2004, 84)).

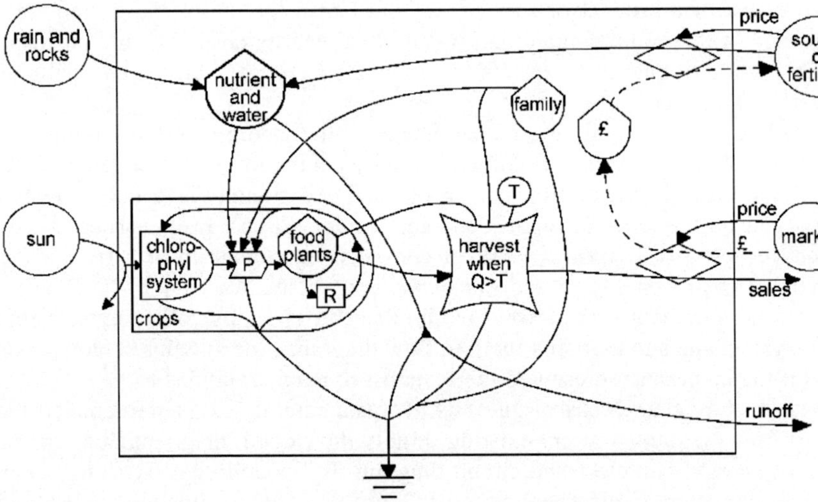

Fig. 6.4 Example of an Energy Systems Language diagram. Source: adapted from Odum (1983, 9).

In this chapter, we have largely stressed the capacity of ESL to model ecosystems. This is appropriate given Odum's background and interests; and most of his work and that of his students and colleagues did indeed use the ESL to study ecosystems rather than other forms. Moreover, he argued that ecosystems are a good way to approach general systems: 'intermediate in size between the microscopic and the astronomical, ecosystems have easily recognizable parts, so that emphasis can be placed on the study of relationships' (Odum 1983, ix). Ecosystems for Odum were not just concerned with the physical environment – the role of humanity as an integral part of ecosystems was critical to his modelling, and he drew out many lessons for agriculture and humanity's relationship with nature from his models.

However, Odum was clear that the modelling language had a much greater capacity, and was relevant to systems of all kinds. The second edition of his book on the ESL was entitled *Ecological and General Systems* (Odum 1994), and indeed Brown (2004) reports that this was Odum's preferred title for the first edition but he was discouraged by his publisher. Odum was very aware of the proliferation of modelling languages in the systems community, and while president of the International Society for the Systems Sciences he 'called for a project to translate models of all scales into systems diagrams so that everyone could better understand them' (Brown et al. 2004, 6). In fact, he went somewhat further than this call for unity, arguing that ESL could become a common diagrammatic form for all systems modelling: "few people can, or often are, given means to read other people's theoretical formulations ... the language could provide a means for eliminating the 'tower of Babel' that now exists in the dispersed literature" (Odum 1983, 579).

It is unfair to regard Odum's call for unity via the ESL as him over-reaching himself – it represented a genuine concern for the multiplicity of modelling approaches and the lack of connections between them. However, a single unified modelling language for systems has yet to be developed, and given the different purposes and approaches to modelling seen in this chapter (and in the whole book) it seems unlikely that such an approach will arise easily.

With energy systems language, we have seen a further example of the transition in systems models from techniques applicable to technological situations (here, electrical circuit diagrams) to human situations; and in this case using largely qualitative approaches through diagrams. In our final example, we will see an even greater emphasis upon diagramming as a way to model systems.

6.5 OU Systems Diagramming: From Engineering to People

Our final case study focuses on an approach that gives primacy to diagrams, treating them as a powerful form of qualitative modelling in their own right (although sitting alongside other forms of modelling). Unlike the other modelling approaches discussed in this chapter, it was developed by a group rather than an individual: the Systems Group at the Open University (OU), UK. Over a period of around thirty years, the OU Systems Group developed a distinctive approach to qualitative modelling that has appeared extensively in the teaching modules of the university, but has been little discussed in wider literature. We present it here not as a form of special pleading for our own university, but because it is a little-told story which gives useful further insights into the trajectories that have occurred within systems modelling.

The Open University is a British distance-learning higher education institution that was founded in 1969. It the largest university in Europe (based on its student numbers, with currently over 200,000 registered students). Almost all of its teaching is carried out at a distance through printed and online course texts, audio-visual materials and various forms of electronic communication. These are sometimes supplemented by face-to-face tuition, through short tutorials or week-long residential schools. Clearly, as media and communication technologies have changed, so have the resources used by the University. For a number of years, it was famous for its late-night broadcasts on the BBC (British Broadcasting Corporation), which have now largely ceased; by contrast the use of computer-based learning has grown steadily over the past thirty years or so.

A 'course' at the OU (which would be termed a module elsewhere) is typically a large entity, requiring 300–600 hours of study, consisting of specially-prepared texts as well as audio-visual material. Such a course is produced by a large team, consisting of several academic authors, as well as media specialists, pedagogical advisors, specialist administrators and others. It is produced with great care, often taking at least two to three years with an expected lifespan of around eight years before replacement (although usually with interim modifications). These courses are delivered by a nationwide network of specialist group tutors. The long lifespan in particular of these courses makes them fairly stable to examine as historical objects, although a deeper such study is beyond the scope of this chapter.

Systems teaching has been part of the Open University since its inception. The Faculty of Technology was one of the first areas to be established, consisting of three disciplines of analysis (electronics, mechanical engineering and materials science) and two disciplines of synthesis (systems and design), although inevitably departmental structures have changed since. The systems group presented its first course in 1972, *Systems Behaviour* (course code T241), which analysed the structure and dynamics of eight different forms of real-world systems: deep-sea container ports, air traffic control, industrial social systems, local government, the British telephone system, ecosystems, the human respiratory system, economic systems, and a shipbuilding firm. Following T241, several different courses were produced which similarly aimed to model a range of different types of systems, as well as many more courses which drew on systems techniques in the teaching of a specific domain (such as the environment, management, or information systems).

To teach systems at a distance, with little if any direct communication between teacher and learner, leads to a quite different approach to systems. Approaches which rely on an apprenticeship model of learning (with a direct relationship between the teacher and an expert in the approach) will not work, nor will techniques which are primarily hands-on. Moreover, the large student numbers (the biggest OU systems course, T301 *Complexity, Management and Change*, had more than 1000 students per year for several years of its life) and long lifespan means that teaching must be quite robust to multiple interpretations. Furthermore, in the early days of the OU, there was no recourse to computer technology; and students were very often working in isolation from others for the bulk of their studies (albeit with occasional tutorial support). These led the group towards the extensive use of diagrams in their teaching, as a way not only to illustrate and help the student to appreciate the nature of systems but to model them and draw conclusions about them.

As Lane and Morris (2001, 719) have written, in one of the few discussions of OU systems diagramming in the wider literature, 'it is our belief that diagrams are an important aid to the process of systems thinking and practice, beyond the need just to communicate particular facts, ideas, or concepts to others in a particular visual product'. Carrying out a brief analysis of key systems books and articles in one of the core journals in the field, they argue that this view is different from that frequently found in the field, where diagrams 'are largely illustrative and their construction not necessarily explained in any detail' (Lane & Morris 2001, 719), although they note the extensive use of diagrams in Checkland and Scholes (1990) and in Senge (1990), two widely-read books in the application of systems thinking to management.

OU systems diagrams developed in form and use over thirty years, culminating in a common text which detailed 25 different diagram types that had been used in various courses (Lane 1999) with guidance on where and how to use each type of diagram. The sources of these diagrams were many. Lane (1999) classifies them into three main types: 'diagrams for exploring complex unbounded situations', 'diagrams for exploring bounded complex situations', and 'diagrams for helping to understand particular structures or processes in a bounded situation' (p.4).

These diagram types were mostly adapted from different kinds of engineering diagrams, to illustrate the structure and behaviour of human-created artefacts. As seen in the list of case studies taught in the course T241 in 1972, many of the systems described in the early courses were precisely this kind of technological system, albeit considered from a sociotechnical viewpoint that looks at the interactions between technologies, people, organisations and society. By the time of the most recent undergraduate course produced by the OU Systems Group, *Understanding systems: making sense of complexity* (T214), produced in 2008, the same case study approach was used but largely to focus on human and social systems: the Internet, the environment, organisations, and criminal justice. Thus we again can see the transition from an engineering to a human perspective common to all the modelling approaches found in this chapter.

Of the 25 diagram types found in OU courses, six have been frequently seen as most significant, and we will discuss each of these in slightly more detail. Each type of diagram was taught in a multimedia package (using Flash) produced by one of us in support of the diagramming pack (Shipp 2002). The six types of diagram are as follows:

- *spray diagrams*: a way of representing a body of connected ideas in a non-linear manner, by beginning with a central term and drawing lines which connect it to a number of sub-terms, which in turn 'spray' out to further sub-terms. They were originally developed by Buzan (1974), and are closely related to his later mind-maps, but are simpler in form.
- *rich pictures*: these are unstructured pictures, usually hand-drawn, of all the major aspects of a situation of interest (see Fig. 6.5). They aim to capture the

full extent of the issues in a situation, before any thought is given to grouping those issues or presenting their structure, although some aspects of the relationships between the issues is frequently illustrated through relative positions or simple lines to connect the issues. Rich pictures were developed by Peter Checkland for the first stage of his Soft Systems Methodology, as a way of exploring the nature of a 'problem situation' (Checkland 1981, 317).

Fig. 6.5 A rich picture about rich pictures.

- *systems maps*: a simple way to illustrate the *structure* of a system of interest. The major components of a system are drawn, either as separate elements or within sub-systems. A key decision in drawing a systems map is deciding what should be within the boundary of the system, and what should be outside of it (in the system's environment). OU systems teaching has long recognised that this decision is a partial one that depends on the perspective of the individual or group drawing the map, and that a different analyst working at a different time might create a very different map. Systems maps derive from engineering diagrams intended to show the structure and components of a human-created system.
- *influence diagrams*: an extension of systems maps to show the relationships between the components within the system (see Fig. 6.6). They closely resemble the entity-relationship diagrams that have long been used within software engineering. By illustrating the influences between components, they enable the analyst to see clearly problems and possibilities arising from those relationships.

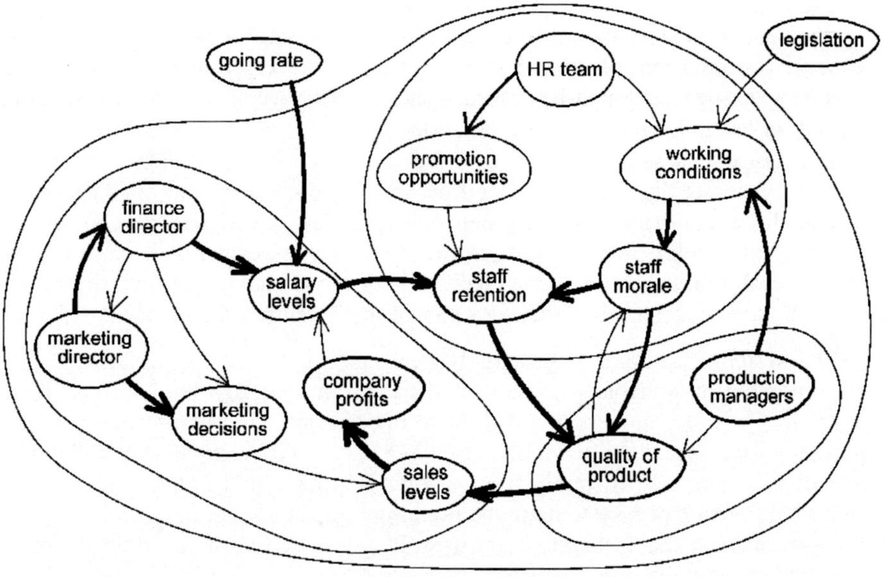

Fig. 6.6 Influence diagram showing some of the influences on company profits in a small production company.

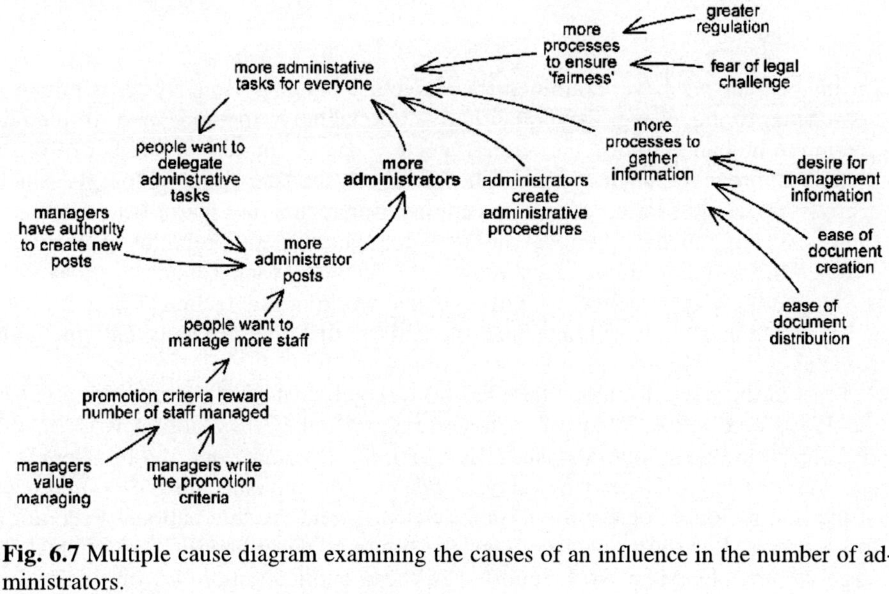

Fig. 6.7 Multiple cause diagram examining the causes of an influence in the number of administrators.

- *multiple cause diagrams*: these rather distinctive diagrams are used to examine *process*, and to make sense of the underlying causes of events (see Fig. 6.7). The purpose of the diagrams is to examine the multiple causes behind particular events and processes, rather than focusing on circular causality in the way that cybernetics does (although they do sometimes contain feedback loops). They serve a similar purpose to the 'fishbone' diagrams widely used in quality management, and developed by Kaoru Ishikawa (1990), but in a free-form manner and for use in a wider range of situations.

- *sign graphs*: these are a specialised form of the causal loop diagrams from system dynamics that were described earlier in the chapter, and emphasise feedback loops and the relationship between variables. In OU systems courses, they are often developed from multiple cause diagrams. Despite their close parallels to causal loop diagrams, Lane (1999, 71) observes that they were 'first used in the biological sciences in the early part of the Twentieth Century'.

OU systems diagrams have largely been used for teaching purposes, but the large number of students who have studied systems through the OU (and British academics at other universities who have tutored on OU systems courses) means that they have spread quite widely, and have been used informally by a number of consultants. The lack of publication of the OU diagrams beyond teaching materials, and the lack of a systematic methodology for their creation, means they have little presence in the academic literature. However, they form a further interesting example of the transition of systems modelling from engineering design to human issues, and towards qualitative approaches.

6.6 Systems Thinking Today: Engineering and Human Systems Together

In this chapter we have examined the trajectory taken by four systems thinking approaches to modelling: from quantitative to qualitative methods, and from engineering to human systems. The same trajectory can be found in a number of other systems approaches, such as Peter Checkland's soft systems methodology, which arose from the application of systems engineering techniques learnt from his experience working in the chemical industry to management problems (Checkland 2000). This common trajectory somewhat reflects the backgrounds of those who developed these approaches, who often began working in a technical field but had a broad humanistic worldview and recognised that their models and methods could assist in wider social and organisational problems.

Most of the work discussed here has been largely historical, being carried out in the 1970s and 1980s. What of systems thinking today? Looking at work in the significant systems journals, such as *Systemic Practice and Action Research* and *Systems Research and Behavioral Science*, or published in the conferences of the International Society for Systems Sciences and the International Federation for Systems Research, we see that the journey described in this chapter is largely complete. Most work reported in these publications relies on qualitative modelling, and they are mostly concerned with human situations, often at the organisational or group level, but increasingly (with environmental concerns being prominent) at a larger scale. The same phenomenon can be seen in the authors discussed in Ramage and Shipp (2009) and in book-length works by contemporary systems thinkers such as Jackson (2003), Bateson (2004) and Meadows (2008).

Not all systems thinkers are so devoted to qualitative approaches. There remains a strong quantitative element in system dynamics, with many articles in the journal *System Dynamics Review* drawing on quantitative models, and strong work in textbook form with a quantitative slant such as Sterman (2000).

The quantitative tradition in systems modelling also continues in a number of areas which have their roots in general systems theory and cybernetics, but which have worked under other labels than 'systems thinking'. Notable here are fields such as complexity science (Kauffman 1995), network theory (Barabasi 2002), systems biology (Werner 2007), and systems engineering (INCOSE 2006). All four of these fields take a 'systemic' approach, in that they are concerned with the behaviour of whole systems, and all four of them are showing noticeable growth, with active conferences and journals in each field. The double trajectory described for systems thinking is partially occurring in a number of these fields, at least in that their area of interest is widening to human situations.

So what lessons can be learnt from the history of modelling within systems thinking that can be applied elsewhere? First, at the start of this chapter we discussed systems thinking's distinctive understanding of the nature of modelling. This understanding is useful to a range of modelling techniques. As we argued earlier, many approaches would advocate such an approach but in practice do not display it.

The second lesson concerns the nature of mid-sized human domains. We have presented the quantitative-qualitative shift as somewhat separate from the engineering-human shift. In fact, one could argue that they are closely linked: as the domain of application has shifted, so too have the modelling techniques useful to making sense of that domain, especially if there is an attempt to create the models in a participatory manner. If other modelling techniques are used in human situations, the same shift towards the qualitative may well occur.

There is a third lesson to be drawn out of the discussion here: that the characteristics of the modeller are often as important as the nuances of the modelling language used. In three of the cases discussed, the modelling techniques arose from the work of a single individual (Jay Forrester, Stafford Beer, and HT Odum), and their success was strongly connected with the abilities of that individual. While the techniques could be taught to others, and in the case of system dynamics in particular with great success, the founder of the approach was able to create models with an ease and dexterity that was much harder for those who followed him. All three of these individuals were colourful and interesting characters, much loved by their students and colleagues, and in many ways their impact came from their personality and exceptional ability as much as from their techniques.

The final lesson from this chapter also concerns the context of the models used. We have somewhat conflated modelling techniques (or language) and the *particular* models created through the use of those techniques. The former may or may not have built-in assumptions. The specific models created very clearly do have assumptions. If we consider environmental or economic modelling, there are huge numbers of possible variables that can be included in the model, and each can take several different starting values. How these variables are selected, and their starting values assigned, is a matter of choice by the modeller(s). The choices involved arise from their assumptions about the nature of the world. In some kinds of models, they might even be called political or ideological choices.

Models are not neutral, objective, statements of reality. They are messy, selective and deeply personal. Good modellers in all fields know this, but by looking at modelling within systems thinking it is especially exemplified and made clear. Models are powerful and interesting, but they are a human artefact, created for a particular purpose by particular people, and it is important to understand this in using and discussing models.

References

Ashby, W.R.: An introduction to cybernetics. Chapman & Hall, London (1956)
Ashby, W.R.: Design for a brain: The origin of adaptive behaviour, 2nd edn. Chapman & Hall, London (1960)
Barabasi, A.-L.: Linked: the new science of networks. Perseus, Cambridge (2002)
Bateson, G.: Steps to an ecology of mind. Chandler, Toronto (1972)
Bateson, M.C.: Willing to learn: Passages of personal discovery. Steerforth Press, Hanover (2004)
Beer, S.: Cybernetics and Management. English Universities Press, London (1959)
Beer, S.: Designing freedom. John Wiley, Chichester (1974)
Beer, S.: Brain of the firm. John Wiley, Chichester (1981)
Beer, S.: The viable system model: Its provenance, development, methodology and pathology. Journal of the Operational Research Society 35(1), 7–26 (1984)
Beer, S.: Diagnosing the system for organizations. John Wiley, Chichester (1985)
Brown, M.T.: A picture is worth a thousand words: energy systems language and simulation. Ecological Modelling 178(1/2), 83–100 (2004)
Brown, M.T., Hall, C.A.S., Jørgensen, S.E.: Eulogy. Ecological Modelling 178(1/2), 1–10 (2004)
Buzan, A.: Use Your Head. BBC Publications, London (1974)
Checkland, P.B.: Systems thinking, systems practice. John Wiley, Chichester (1981)
Checkland, P.B.: Soft systems methodology: A thirty year retrospective. Systems Research and Behavioral Science 17(S1), S11–S58 (2000)
Checkland, P.B., Scholes, J.: Soft systems methodology in action. John Wiley, Chichester (1990)
Forrester, J.W.: Industrial dynamics: A major breakthrough for decision makers. Harvard Business Review 36(4), 37–66 (1958)
Forrester, J.W.: Industrial dynamics. MIT Press, Cambridge (1961)
Forrester, J.W.: The beginning of system dynamics. System Dynamics Society, Stuttgart (1989), http://www.clexchange.org/ftp/documents/system-dynamics/SD1989-07BeginningofSD.pdf (accessed July 11, 2011)
Forrester, J.W.: System dynamics – a personal view of the first fifty years. System Dynamics Review 23(2/3), 345–358 (2007)
Hardin, G.: The Tragedy of the Commons. Science 162(3859), 1243–1248 (1968)
Hoverstadt, P.: The Viable System Model. In: Reynolds, M., Holwell, S. (eds.) Systems Approaches to Managing Change: A Practical Guide, pp. 87–133. Springer, London (2010)

INCOSE, A Consensus of the INCOSE Fellows. International Council on Systems Engineering (2006), http://www.incose.org/practice/fellowsconsensus.aspx (accessed July 11, 2011)
Ishikawa, K.: Introduction to Quality Control. Chapman and Hall, London (1990)
Jackson, M.C.: Evaluating the managerial significance of the VSM. In: Espejo, R., Harnden, R. (eds.) The Viable System Model Revisited: Interpretations and Applications of Stafford Beer's VSM, pp. 407–439. John Wiley, Chichester (1989)
Jackson, M.C.: Systems thinking: Creative holism for managers. John Wiley, Chichester (2003)
Kauffman, S.: At home in the universe: The search for the laws of self-organization and complexity. Penguin, London (1995)
Lane, A.: Systems Thinking and Practice: Diagramming. The Open University, Milton Keynes (1999)
Lane, A., Morris, D.: Teaching Diagramming at a Distance: Seeing the Human Wood Through the Technological Trees. Systemic Practice and Action Research 14(6), 715–734 (2001)
Lane, D.C.: Trying to think systematically about 'Systems Thinking'. Journal of the Operational Research Society 46(9), 1158–1162 (1995)
Maani, K.E., Cavana, R.Y.: Systems Thinking and Modelling: Understanding Change and Complexity. Pearson Education, Auckland (2000)
Meadows, D.H.: Thinking in Systems: A Primer. In: Wright, D. (ed.) White River Junction. Chelsea Green Publishing, VT (2008)
Meadows, D.H., Meadows, D.L., Randers, J., Behrens, W.W.: The limits to growth: A report for the club of Rome's project on the predicament of mankind. Universe Books, New York (1972)
Morris, D., Chapman, J.: Systems Thinking and Practice: Modelling. The Open University, Milton Keynes (1999)
Nguyen, N.C., Bosch, O.J.H., Maani, K.E.: Creating 'learning laboratories' for sustainable development in biospheres: A systems thinking approach. Systems Research and Behavioral Science 28(1), 51–62 (2011)
Odum, E.P.: Fundamentals of ecology, 3rd edn. W.B. Saunders, Philadelphia (1971)
Odum, H.T.: Trophic structure and productivity of Silver Springs, Florida. Ecological Monographs 27(1), 55–112 (1957)
Odum, H.T.: Systems ecology: An introduction. John Wiley, New York (1983)
Odum, H.T.: Ecological and general systems: an introduction to systems ecology. University Press of Colorado, Niwot (1994)
Pickering, A.: The science of the unknowable: Stafford Beer's cybernetic informatics. Kybernetes 33(3/4), 499–521 (2004)
Pidd, M.: Tools for Thinking: Modelling in Management Science, 2nd edn. John Wiley, Chichester (2003)
Ramage, M., Shipp, K.: Systems Thinkers. Springer, London (2009)
Richardson, G.P.: Foreword. In: Maani, K.E., Cavana, R.Y. (eds.) Systems Thinking and Modelling: Understanding Change and Complexity, pp. vii–viii. Pearson Education, Auckland (2000)
Rosenhead, J.: IFORS operational research hall of fame: Stafford Beer. International Transactions in Operational Research 13(6), 577–581 (2006)
Senge, P.M.: The fifth discipline: the art and practice of the learning organization. Doubleday, New York (1990)

Shipp, K.: Systems Thinking and Practice: Diagramming CD-ROM. The Open University, Milton Keynes (2002)

Sterman, J.D.: Business dynamics: Systems thinking and modeling for a complex world. Irwin/McGraw-Hill, Boston (2000)

Taylor, P.J.: Technocratic Optimism, H. T. Odum, and the Partial Transformation of Ecological Metaphor after World War II. Journal of the History of Biology 21(2), 213–244 (1988)

Taylor, P.J.: Unruly complexity: Ecology, interpretation, engagement. University of Chicago Press, Chicago (2005)

Turner, G.: A comparison of the Limits to Growth with thirty years of reality. Commonwealth Scientific and Industrial Research Organisation, CSIRO (2008), `http://www.csiro.au/files/files/plje.pdf` (accessed July 11, 2011)

Werner, E.: All systems go. Nature 446(7135), 493–494 (2007)

Wiener, N.: Cybernetics: or control and communication in the animal and the machine. MIT Press, Cambridge (1948)

Wolstenholme, E.F.: Qualitative vs quantitative modelling: The evolving balance. Journal of the Operational Research Society 50(4), 422–428 (1999)

Chapter 7
Visualisations for Understanding Complex Economic Systems

Marcel Boumans

University of Amsterdam, The Netherlands

Abstract. In the history of economics, a few (but famous) analogue systems were built with the purpose of gaining a better understanding of an economics mechanism by creating a visualisation of it. One of the first was Irving Fisher's mechanism, constructed in 1893, consisting of a tank with floating cisterns connected by sticks visualising a three-good, three consumer economy. More famous and better-known is the Phillips-Newlyn Hydraulic Machine, built in 1949, representing macroeconomics by flows and stocks of coloured water in a system of Perspex tanks and channels. This hydraulic machine became a reference point for developing other less fragile systems to visualise an economic mechanism, namely by simulations run on a computer. The main part of this chapter will discuss FYSIOEN, a computer visualisation of a hydraulic system representing the macro-econometric model MORKMON of the Dutch Central Bank, designed in 1988. FYSIOEN was developed to help users gain understanding of the complex mathematical model by translating it into the visual domain. An analogy usually transfers a familiar mechanism to an unfamiliar domain in order to provide an understanding of the latter. So, in case of the analogues of Fisher, and Phillips and Newlyn, the more familiar hydraulic laws were used to attain understanding of an economic mechanism. The problem with FYSIOEN, however, was that although it was an animation of a hydraulic system, the program was not run by hydraulic laws but by the relationships used in MORKMON. Several improvements were suggested to make the animation look more real to compensate for the lack of hydraulic laws, but computing facilities at the time limited the possibilities.

7.1 Introduction

In the late 1980s, a 'graphic model', called FYSIOEN (Kramer et al. 1988), was developed at the Dutch central bank, *De Nederlandsche Bank* (DNB), to visualise the internal operation of the Bank's economic model MORKMON:

> Many economic policy models have evolved into highly complex systems which are difficult for outsiders to understand. As such models are often operated on mainframes,

C. Bissell and C. Dillon (Eds.): Ways of Thinking, Ways of Seeing, ACES 1, pp. 145–165.
springerlink.com

> access is limited. Moreover, output is usually confined to the results produced by the model, thus affording no insight into the manner in which they were achieved. This may either undermine confidence in the results or may lend them an aura of sacrosanctity. Both these outcomes are undesirable. (Kramer et al. 1990, 159)

FYSIOEN, a Dutch acronym for physical-visual operational economic model for the Netherlands[1], was meant as a 'didactic tool': 'The visual model affords more rapid and better insight – to more advanced model users as well – than a set of mathematical equations with explanatory notes' (p. 159). MORKMON is a 164-equation econometric model, meaning that the mathematical model has had its parameters estimated using statistical data and techniques, which as such is not an intelligible model.

To arrive at an intelligible visualisation of MORKMON, FYSIOEN was designed to describe the economy on the basis of analogies between hydraulics and economics: 'The human being appears to be capable of recognising the principles of one system in another system, where this recognition can lead to understanding' (Mosselman 1987, 3, trans.[2]).

The exemplar model was the Phillips-Newlyn machine (Phillips 1950), a hydraulic model (7×5×3 ft, or 2.1×1.5×0.9 m) representing the macroeconomy by flows and stocks of coloured water in a system of Perspex tanks and channels. A small number of these machines was made and these found their way across the world. They were used for demonstration and teaching in the 1950s and early 1960s. Trained as an electrical engineer and having started a degree at the London School of Economics after World War II, Bill Phillips (1914–1975) had difficulties with understanding the macroeconomics of those days. In Britain, macroeconomics was based around mainly verbal elucidations and extensions of the ideas found in John Maynard Keynes' *General Theory* (1936). To get a grip on this macroeconomics thinking, and to resolve his own difficulties in understanding, Phillips used his engineering skills to create together with Walter Newlyn (1915–2002), a lecturer in economics at Leeds University, UK, the famous (in economics, at least) hydraulic machine.[3]

FYSIOEN is, however, not a physical model but a computer animation of a hydraulic system representing MORKMON. The animation did not work according to the laws of hydraulics but according to the equations of the econometric model. The equations guiding the motion of the pictures were translated by a graphics package which interpreted them in terms of colours and sizes of predetermined shapes.

> [T]he design of all images is based on an imaginary system of tubes and basins. Flows pass through the tubes, changing the liquid levels in the basins, i.e. the volume of the stock variables. Relations between variables are reflected in merging or branching flows. Economic forces are shown as taps, balance beams, springs, pumps, pistons and handles.

---

[1] FYSIek-visueel Operationeel Economisch model voor Nederland. FYSIOEN is phonetically equal to *visioen*, the Dutch term for vision.

[2] All quotations from Dutch texts are translated by the author. Each time this is indicated by 'trans.'.

[3] Morgan and Boumans (2004) provide a detailed account of this model and its making. See also Bissell (2007).

> Links between these devices also illustrate the relationships within each image. Floats, pressure gauges and taps record and control the volume of flows and stocks. (Kramer et al. 1990, 153)

The unfamiliar mathematics was translated into an animated representation that seemed to be in the more familiar language of hydraulics, yet it did not work according to hydraulic laws. It appeared that the animation was an illusion: it might happen in the FYSIOEN animation that a cistern fills up without a tap being opened. If a visualisation contradicts our experience, it will lead to confusion, disorientation, or astonishment. Sometimes this is for aesthetic reasons, like the optical illusions in Escher's etchings, or sometimes just for fun, like the Tom and Jerry cartoons. But when these illusionary things happen it does not lead to a better understanding of either hydraulic or economic principles.

An analysis of the development of FYSIOEN shows us the conditions for which a visualisation leads to an understanding of an unintelligible system (Sect. 7.2). The visualisation in case of FYSIOEN is based on analogous modelling. An analogy is the transference of a familiar conceptual framework appropriate to one subject to another, unfamiliar subject, in order to gain a better understanding of the latter. The question is what should be transferred to gain understanding. To answer this question, Sect. 7.3 presents the history of models which is intrinsically interwoven with the history of the method of analogy and the history of intelligibility. This history shows that to gain understanding, the familiar essential relationships should be transferred. This is however a necessary condition for understanding not a sufficient one. In the 1950s, electronic analogues were built to investigate economic systems (Sect. 7.4). Although they were analogies transferring relationships, they were not intelligible because they did not enable a visualisation of the process ruled by these relationships but only the process's outcomes. To gain understanding of a complex system, its dynamic characteristics should be transferred to the visual and familiar domain, for which hydraulic systems indeed seem to be most appropriate.

7.2 Visualisation of MORKMON

The development of FYSOEN started as a 'project to design a physical model of MORKMON' at the beginning of 1985. As the project's title indicates, the original goal was to design a physical model of MORKMON, called F-MORKMON (Kramer and Coenen 1985a). The econometric model MORKMON was called E-MORKMON. This project took the following considerations as a starting point for developing F-MORKMON:

- *Analogies* between economic and physical systems should only be considered with respect to the mutual use of the same logical and analytical concepts.
- Various different kinds of systems (hydraulic, mechanical, thermal, biological, chemical) can be described by *systems of equations.* A hydraulic version of MORKMON is only just one of the possibilities. The only appeal of a hydraulic model above other systems is that the liquid flow is visible, and it shows similarity with our image of commodity and money flows (circular flow).

- A physical model cannot be more than an *illustration* of a number of basic concepts from MORKMON. Equilibrium, causality, time and elasticity, for example, can be depicted. 'Growth' is much more difficult. (Kramer and Coenen 1985a, 1–2, trans.).

The reference model for designing F-MORKMON was the Phillips machine.[4] Kramer and Coenen (1985a) suggested that the workings and performance of the Phillips machine could be improved by using microelectronics:

- To control the physical model, as in industrial processes such as water and gas distribution, and chemical processes (refinery or brewery).
- To support the presentation of the results by using graphical terminals, screens, video, printer and so on. (p.7, trans.).

The considerations of applying microelectronics in both ways (control and support) were:

- A computer program of MORKMON was available; the simulation results therefore could be made visible on a graphic terminal.
- Simple interactive control was aimed at; the Phillips machine was not user-friendly at all.
- Improving accuracy. (p. 7, trans.).

It was considered a disadvantage, however, that electronic control as such was not as visual as the hydraulic Phillips machine.

These lists of considerations resulted in the following requirements: The physical model should be an appropriate 'illustration' of the econometric model, that is F-MORKMON must be a 'recognisable reflection' of E-MORKMON. Stylisation of E-MORKMON will be necessary. Analytical concepts such as equilibrium, causality, time and measure have to be represented clearly. Interactive control has to be simple and insensitive to disturbances.

The idea of an improved version of the Phillips-Newlyn hydraulic system using microelectronics gradually met some opposition at the Bank. It would require some specific measures with respect to its location. In a note on the progress of the project, it was explicitly mentioned that a hydraulic model could not be placed in the high-rise building of the Bank (Kramer and Coenen 1985b). Moreover, a hydraulic system like the Phillips-Newlyn machine would require too much maintenance, and starting up a demonstration with it was known to be quite problematic (see also Morgan and Boumans 2004). Therefore, DNB approached the Control Engineering Laboratory of the Delft University of Technology, represented by H.R. van Nauta Lemke and P.P. J. van den Bosch. The resulting cooperation focussed on the 'visualisation of MORKMON', to be carried out as a master thesis project by a student L.J. Dingemanse. At this stage of the project, the Bank was still aiming at a physical model for demonstrations. The target group for demonstration was not only economists but also a wider audience (Bikker and Kramer 1986).

[4] Newlyn's share in the development of this hydraulic machine became gradually ignored in the history of economics.

To speed up the development of a visualisation of MORKMON, Dingemanse was asked to work on a stylised version of MORKMON. The stylisation meant primarily a reduction of the model variables: from 220 to 70 endogenous variables and from 175 to 60 exogenous variables[5] (Kramer and Bikker 1986a). This stylised version of MORKMON became quite soon to be called MINIMORKMON (Kramer and Bikker 1986b), in which model reduction was slightly less: from 220 to 80 endogenous variables and from 175 to 70 exogenous variables.

The visualisation of MORKMON by Dingemanse (1987) was not a physical hydraulic model, like the Phillips model, controlled by a computer, but rather a visualised hydraulic analogy on a screen. Although Dingemanse noted that in a later stage a computer-controlled physical model could still be developed, it never actually happened.

Dingemanse listed four practical disadvantages of a hydraulic physical model:

- Only experiments within a very restricted area are possible. The pipes and tanks fix the experimental space.
- Maintenance and operation are very difficult. It is difficult to arrive at a specific equilibrium position due to algal growth and scale.
- Not very flexible. The magnitudes and their dimensions are fixed by the structure of the model, and not easily changed.
- Very difficult to transport. (pp. 3–4, trans.)

Notwithstanding these disadvantages, Dingemanse had investigated and compared various alternative physical analogies: hydraulic, pneumatic, thermal, electronic, and mechanical, and concluded that: 'Considering the illustrative power of streaming water as analogy of money flows and also because of practical considerations, a hydraulic model is the best solution for constructing a physical analogue model of MORKMON' (p. 7, trans.):

- In the case of an electronic analogue, the interactions between the electronic magnitudes, such as current and voltage, cannot be made visible, only their results.
- The construction of a thermal analogue of considerable size is practically impossible.
- A mechanical analogue can be a visualisation. Its construction would be an assemblage of mechanical calculators such as integrators, multipliers and adders. The problem, however, is that it requires some level of mechanical knowledge.
- A pneumatic model has to be closed. Changes of volume can be made visible by the movement of pistons, but air flows cannot easily be made visible. One could add small particles to the air to indicate the flows, but that would make it even more difficult to describe and to control the model's behaviour.
- In contrast to the above physical models, in a hydraulic model 'the flows are visible by hearing the noise and seeing the circumference of the spout' (Dingemanse, 1987, 8, trans.).

[5] An endogenous variable is a variable that is determined by one or more variables in the model. An exogenous variable is not determined by any of the model variables.

To develop a hydraulic model, MORKMON would have to be simplified too much, and the result would not be flexible enough. A computer-controlled physical model could be seen as an extension of a visualisation of a physical model on screen, therefore it was decided to focus on the latter option: a visualisation on screen. To speed up the project, it was decided to work with a simplified version of MORKMON, now called MINIMO.

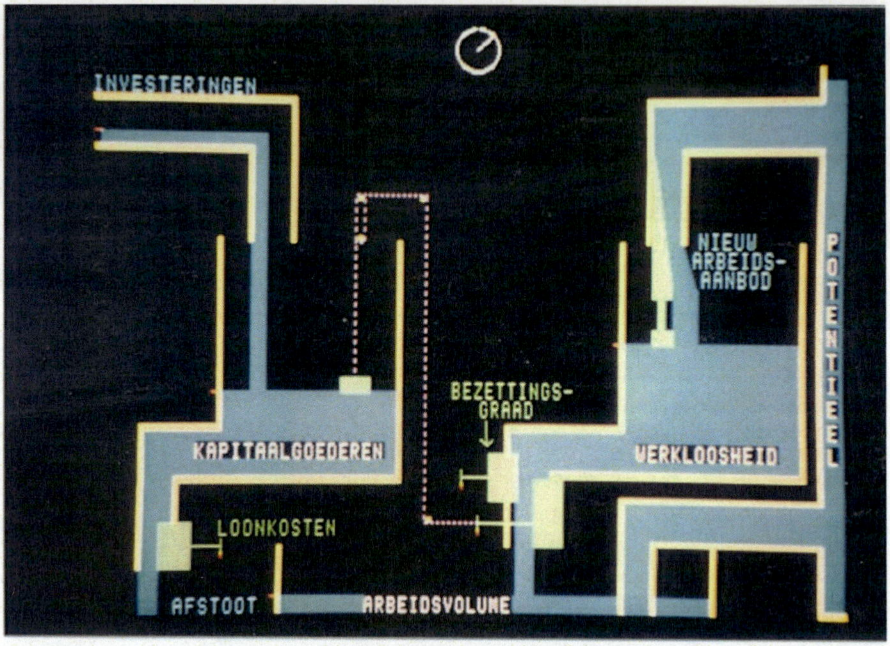

Fig. 7.1 Visualised hydraulic analogy. Source: Dingemanse (1987, 24).

A test version, a prototype, of a visualised hydraulic analogy on screen was developed, see Fig. 7.1. Water is expected to move due to gravity. Where the water rises a red circle in a tube indicates that the water is being pumped up.

> The visualised model on a screen is most intelligible if it is based on hydraulic analogies, supplied with mechanical elements. This comes closest to the experience and foreknowledge of the audience. (Dingemanse 1987, 51, trans.)

At the end of 1986, Kramer (1986) summarised the results of the project in a note in which he gave F-MORKMON its current name, FYSIOEN, emphasising the shift from a physical model to a visualisation on screen of a physical model, from the real to the virtual world.

In December 1986 and January 1987 two test demonstrations were organised with different kinds of audiences. Based on the responses recorded at these two demonstrations, J.A. Reus (1987) developed suggestions for the improvement of the MORKMON-visualisation. One of the main points of critique was that the visualisation was not lively enough. To make the visualisation 'look more like a real physical analogy', Reus suggested adding noises of streaming water, and sounds of moving stopcocks and of running pumps (p. 41). Other suggestions were to add air bubbles to the water to visualise the streaming of the water, or to replace the smooth water levels by small moving waves. None of these suggestions were ever implemented.

Instead of making the visualisation more lively, J.F. Mosselman (1987) was put to work on making the simulation more interactive, another point of critique that came up at the test demonstrations. This result was demonstrated at the Economic Modelling Conference held in Amsterdam in October 1987.

Shortly after this event, FYSIOEN (1988) was published: a monograph including two PC diskettes, a 3.5 inch diskette for an IBM compatible PS/2 model 50 and higher, and a 5.25 inch diskette for an IBM compatible PC, XT and AT.

7.3 Models as Analogies

The tradition of seeing models as analogies is rooted in the work by James Clerk Maxwell (1831–1879). In Maxwell's work, a heuristic shift took place that was to lead to a new method of modern physics.

In his first paper on electromagnetism, 'On Faraday's lines of force' (1855/1965) (see Boltzmann 1892/1974), Maxwell set out the method he intended to use. He suggested that to study effectively the considerable body of results from previous investigations, the results have to be simplified and reduced to 'a form in which the mind can grasp them'. On the one hand they could take the form of 'a purely mathematical formula', but then one would 'entirely lose sight of the phenomena to be explained' (Maxwell 1855/1965, 155). On the other hand, if they were to take the form of a 'physical hypothesis', that is, an assumption as to the real nature of the phenomena to be explained, this would mean that 'we see the phenomena only through a medium', making us 'liable to that blindness to facts and rashness in assumptions which a partial explanation encourages' (pp. 156–6):

> We must therefore discover some method of investigation which allows the mind at every step to lay hold of a clear physical conception, without being committed to any theory founded on the physical science from which that conception is borrowed, so that it is neither drawn aside from the subject in pursuit of analytical subtleties, nor carried beyond the truth by a favourite hypothesis. (Maxwell 1855/1965, 156)

To obtain physical ideas without adopting a physical theory we have to exploit 'dynamical analogies', 'that partial similarity between the laws of one science and those of another which makes each of them illustrate the other' (p. 156). In other words, to the extent that two physical systems obey laws with the same mathematical form, the behaviour of one system can be understood by studying the behaviour of the other, better known, system. Moreover, this can be done without making any hypothesis about the real nature of the system under investigation.

In a later paper, 'On the mathematical classification of physical quantities' (1871/1965), Maxwell drew a distinction between a 'physical analogy' and a 'mathematical or formal analogy'. In the case of a formal analogy, 'we learn that a certain system of quantities in a new science stand to one another in the same mathematical relations as a certain other system in an old science, which has already been reduced to a mathematical form, and its problems solved by mathematicians' (pp. 257–8). We can speak of a physical analogy when, in addition to a mathematical analogy between two physical systems, we can identify the entities or properties of both systems.

Maxwell's distinction between these two kinds of analogies was expounded in more detail by Ernest Nagel (1961). In physical analogies, which he called 'substantive analogies':

> a system of elements possessing certain already familiar properties, assumed to be related in known ways as stated in a set of laws for the system, is taken as a model for the construction of a theory for some second system. This second system may differ from the initial one only in containing a more inclusive set of elements, all of which have properties entirely similar to those in the model; or the second system may differ from the initial one in a more radical manner, in that the elements constituting it have properties not found in the model (or at any rate not mentioned in the stated laws for the model). (Nagel 1961, 110)

In formal analogies:

> the system that serves as the model for constructing a theory is some familiar structure of abstract relations, rather than, as in substantive analogies, a more or less visualizable set of elements which stand to each other in familiar relations. (Nagel 1961, 110)

For Heinrich Hertz (1857–1894), representations of mechanical phenomena could only be understood in the sense of Maxwell's dynamical analogies, which is obvious in the section 'Dynamical models' of his last work, *The Principles of Mechanics Presented in a New Form* (1899/1956). First he gave a definition of a 'dynamical model':

> A material system is said to be a dynamical model of the second system when the connections of the first can be expressed by such coordinates as to satisfy the following conditions:
>
> 1. That the number of coordinates of the first system is equal to the number of the second.
> 2. That with a suitable arrangement of the coordinates for both systems the same equations of condition exist.
> 3. That by this arrangement of the coordinates the expression for the magnitude of a displacement agrees in both systems. (Hertz 1899/1956, 175)

From this definition, Hertz inferred that 'In order to determine beforehand the course of the natural motion of a material system, it is sufficient to have a model of that system. The model may be much simpler than the system whose motion it represents' (p. 176). However,

> It is impossible to carry our knowledge of the connections of the natural systems further than is involved in specifying models of the actual systems. We can then, in fact, have no knowledge as to whether the systems which we consider in mechanics agree in any other respect with the actual system of nature which we intend to consider, than this alone, - that the one set of equations are models of the other. (Hertz 1899/1956, 177)

While the 'model' was still considered as something material, a 3-dimensional object, its relationship to the system of inquiry should be the same as the relationship of the images (*Bilder*) we make of the system to the system itself; namely, that the consequents of the representation, whether material (model) or immaterial (image), must be the representation of the consequents. However, this relationship between a representation and the system under investigation would allow for many different representations. Hertz, therefore, formulated three requirements a representation should fulfil. First, a representation should be 'logically permissible', that is, it should not contradict the principles of logic. Second, permissible representations should be 'correct', that is, the relations of the representation should not contradict the system relations. Third, of two correct and permissible representations of the same system, one should choose the most 'appropriate'. A representation is more appropriate when it is more distinct, that is, when it contains more of the essential relations of the system; and when it is simpler, that is, when it contains a smaller number of superfluous or 'empty' relations. Hertz explicitly noted that empty relations cannot be altogether avoided: 'They enter into the images because they are simply images, - images produced by our mind and necessarily affected by the characteristics of its mode of portrayal' (p. 2).

In short, the three requirements that a representation of a system should fulfil are: (1) logical consistency; (2) 'correctness', that there is correspondence between the relations of the representation and those of the system; and (3) 'appropriateness', that is contains the essential characteristics of the system as simply as possible.

Like Hertz, Ludwig Boltzmann (1844–1906) placed great importance on Maxwell's concept of analogies, describing Maxwell as having been 'as much of a pioneer in epistemology as in theoretical physics' (Boltzmann 1912, 100):

> Most surprising and far-reaching analogies revealed themselves between apparently quite disparate natural processes. It seemed that nature had built the most various things on exactly the same pattern; or, in the dry words of the analyst, the same differential equations hold for the most various phenomena. (Boltzmann 1892/1974, 9)

According to Boltzmann (1902b/1974), 149), 'It is the ubiquitous task of science to explain the more complex in terms of the simpler; or, if preferred, to represent the complex by means of a clear picture borrowed from the sphere of the simpler phenomena'. Boltzmann's attitude towards the role of '*Bilder*' in physics was explicitly expressed in an essay 'On the development of the methods of theoretical physics' (1899a/1974). Referring to the Hertz 'programme', Boltzmann stated that:

> no theory can be objective, actually coinciding with nature, but rather that each theory is only a mental picture of phenomena, related to them as *sign* is to *designatum*. From this it follows that it cannot be our task to find an absolutely correct theory but rather a picture that is, as simple as possible and that represents phenomena as accurately as possible. One might even conceive of two quite different theories both equally simple and equally congruent with phenomena, which therefore in spite of their difference are equally correct. (Boltzmann 1899a/1974, 90–91)

Although Boltzmann frequently referred to Hertz when discussing '*Bilder*' there is an important difference between the two men (see De Regt (1999, 116)). Boltzmann rejected Hertz's first requirement that the picture we construct must obey the principles of logic as 'indubitably incorrect': 'the sole and final decision as to whether the picture is appropriate lies in the circumstance that they represent experience simply and appropriately throughout so that this in turn provides precisely the test for the correctness of those laws' (Boltzmann 1899b/1974, 105).

In mathematics and physics, the term 'model' originally referred specifically to material objects (see for example Hertz's definition of a 'dynamical model' above), 'a representation in three dimensions of some projected or existing structure, or of some material object artificial or natural, showing the proportions and arrangement of its component parts', or 'an object or figure in clay, wax, or the like, and intended to be reproduced in a more durable material' (*Oxford English Dictionary*, 1933). Boltzmann's entry for 'Model' in the *Encyclopaedia Britannica* (1902a /1974, 213) also indicates its materiality: 'a tangible representation, whether the size be equal, or greater, or smaller, of an object which is either in actual existence, or has to be constructed in fact or thought'. Today, 'model' can mean both a material object and an image, 'Bild', in the sense of Hertz.

7.3.1 Understanding by Models

Understanding by models fits into a longer tradition that started with what Galileo (1564–1642) took to be intelligible and the concept of intelligibility that he developed. Machamer (1998) shows that Archimedean simple machines, such as the balance, the inclined plane, and the screw, combined with the experiences gained using them, constituted Galileo's concept of intelligibility:

> Intelligibility or having a true explanation for Galileo had to include having a mechanical model or representation of the phenomenon. In this sense, Galileo added something to the traditional criteria of mathematical description (from the mixed sciences) and observation (from astronomy) for constructing scientific objects (as some would say). ... To get at the true cause, you must replicate or reproduce the effects by constructing an artificial device so that the effects can be seen. (Machamer 1998, 69)

According to Machamer and Woody (1994), Archimedean simple machines are models of intelligibility having the following property: because it has a concrete instantiation in the real world the model can be acted upon and manipulated experimentally. Also because of this it is visualisable or picturable (p. 222).

This mode of scientific understanding was also emphasised by William Thomson (Lord Kelvin, 1824–1907) by his well-known dictum:

> It seems to me that the test of 'Do we or do we not understand a particular subject in physics?' is, 'Can we make a mechanical model of it?' (Thomson 1884/1987, 111)

In this tradition, understanding a phenomenon became the same as 'designing a model imitating the phenomenon; whence the nature of material things is to be understood by imagining a mechanism whose performance will represent and simulate the properties of the bodies' (Duhem 1954, 72).

De Regt (1999) shows how in Boltzmann's philosophy of science *Bilder* functioned as tools for understanding. The kind of images Boltzmann preferred, as being most intelligible, were mechanical pictures: 'it is the practical success of mechanicism – possibly linked with our familiarity with mechanical systems from daily experience – that has made it into a criterion for intelligibility in science':

> What, then, is meant by having perfectly correct understanding of a mechanism? Everybody knows that the practical criterion for this consists in being able to handle it correctly. However, I go further and assert that this is the only tenable definition of understanding a mechanism. (Boltzmann 1902b/1974, 150)

This linkage between understanding by a model and being able to handle the model was rediscovered by Morrison and Morgan (1999). They demonstrate that models function as 'instruments of investigation' helping us to learn more about theories and the real world because they are autonomous agents, that is to say, though they represent either some aspect of the world, or some aspect of a theory, they are partially independent of both theories and the world. It is precisely this partial independency that enables us to learn something about the thing they represent, but:

> We do not learn much from looking at a model – we learn more from building the model and manipulating it. Just as one needs to use or observe the use of a hammer in order to really understand its function, similarly, models have to be used before they will give up their secrets. In this sense, they have the quality of a technology – the power of the model only becomes apparent in the context of its use. (Morrison and Morgan 1999, 12)

7.3.2 Economic Models of Intelligibility

In line with this Galileo-Maxwell-Boltzmann tradition, Irving Fisher (1867–1947), one of the founders of modern economics, was convinced that understanding a certain mechanism or phenomenon demands visualisation, 'for correct visual pictures usually yield the clearest concepts' (Fisher 1939, 311). Sometimes these pictures showed mechanical devices, because he believed that 'a student of economics thinks in terms of mechanics far more than geometry, and a mechanical illustration corresponds more fully to his antecedent notions than a graphical one' (Fisher 1892/1925, 24). Fisher (1892/1925) used pictures of a hydrostatic mechanism to explain a three-good, three consumer economy in his Ph.D. thesis. This mechanism was later built and used in teachings. He also used a balance to illustrate the equation of exchange and a hydraulic system 'to observe and trace' important variations and their effects, in the *Purchasing Power of Money* (Fisher 1911/1963, 108).

7.4 Electronic Analogues

Fisher's 'hydraulic machines' were explicitly referred to in Kramer et al. (1988), including a picture of Fisher's 'price level machine', but not the electronic analogues that were developed and built in the 1950s and used for economic investigations.[6] One of the protagonists of this kind of research was Robert H. Strotz (1922–1994) of the Department of Economics, Northwestern University. Together with John F. Calvert, S.J. Horwitz, James C. McAnulty, Nye Frank Morehouse, and J.B. Naines of the Aerial Measurements Laboratory of the Technical Institute

[6] See (Small 1993) for a more general historical context.

at Northwestern University, he used the Aeracom, which was made available through the cooperation of the Navy Department, Bureau of Aeronautics:[7]

> The Aeracom is a room-sized machine designed for the ready construction of various electrical circuits. Once a particular circuit has been connected, variation in one (or more) of the electrical magnitudes (e.g., voltage, resistance, capacitance, etc.) causes variation in the other variables of the system. The effect on each of the other (dependent) variables may then be seen on an oscilloscope screen as a time series and recorded photographically. Changes in the parameter values are easily made by turning dials. (Morehouse et al. 1950, 314)

The Aeracom was used for the study of physical and engineering problems.

> The mathematical representation of mechanical, acoustic, and hydraulic systems may often be found to correspond to the mathematical representation of contrived electrical circuits. Hence, by analogy, the mathematical solution of the electrical system is the same as that of the mechanical, acoustic, or hydraulic system under investigation. It is the thesis of this paper that electrical analogs may also be constructed for economic models and the properties of the models investigated in a similar way. (Morehouse et al. 1950, 314)

The first paper (Morehouse et al. 1950) used only a 'simple' inventory model to illustrate the applicability of the Aeracom. The problem being investigated was depicted as shown in Fig. 7 2:

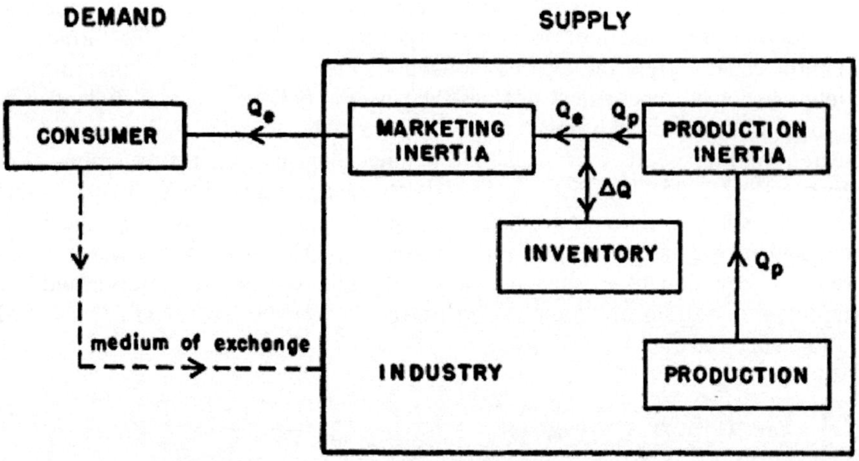

Fig. 7.2 Economic model. Source: Morehouse et al. (1950, 314).

---

[7] Ludington Daily News, September 20, 1950 contains the following information:

> The Northwestern computer is operated under the direction of Dr. John F. Calvert, chairman of the department of electrical engineering and director of the laboratory, and James C. McAnulty, technical director. R.H. Strotz, an instructor of Northwestern, and N.F. Morehouse, a student, already have used the computer to work out intricate theoretical problems in economics.

The model was given by the following equations:

demand function $P_d = \alpha_1 - \beta_1 Q_e$

output function $P_s = \alpha_2 - \beta_2 Q_p$

accelerator principle $P_d - P_s = \lambda_1 \dot{Q}_e + \lambda_2 \dot{Q}_p$

overcompensation principle $P_d - P^0 = \lambda_1 \dot{Q}_e + \frac{1}{\gamma} \int_{T^0}^{T} (Q_e - Q_p) dT + P_i$

adjustment $P_s - P^0 = -\lambda_2 \dot{Q}_p + \frac{1}{\gamma} \int_{T^0}^{T} (Q_e - Q_p) dT + P_i$

where Q_e is the quantity exchanged, P_d the demand price for this quantity, Q_p is the quantity produced, and P_s is the lowest price at which this quantity is produced. T is time and P_i is an inventory control function. The analogous equations in electrical terms were:

$$E_1 = V_1 - R_1 I_1$$

$$E_2 = V_2 - R_2 I_2$$

$$E_1 - E_2 = L_1 \dot{I}_1 + L_2 \dot{I}_2$$

$$E_1 - E^0 = L_1 \dot{I}_1 + \frac{1}{C} \int_{t^0}^{t} (I_1 - I_2) dt + E_i$$

$$E_2 - E^0 = -L_2 \dot{I}_2 + \frac{1}{C} \int_{t^0}^{t} (I_1 - I_2) dt + E_i$$

where R is resistance, L inductance, C capacitance, V battery voltage, E voltage across circuit elements, I current, and t is time.

From these equations a circuit of electrical elements was devised as shown in Fig. 7.3, where q^0 is the initial charge on C, and D_1 and D_2 are diodes permitting current flow only in the direction indicated.

This circuit was 'plugged' into the Aeracom and the values of the elements were set to simulate a reference equilibrium position for the model. Then the equilibrium of the circuit was disturbed by suddenly increasing the value of the voltage V_1, and the behaviour of some selected variable was observed as a time function on the oscilloscope screen. After several different variables had been selected in turn, the values of the elements were rapidly changed and behaviours of the new conditions were obtained and recorded photographically, as shown in Fig. 7.4.

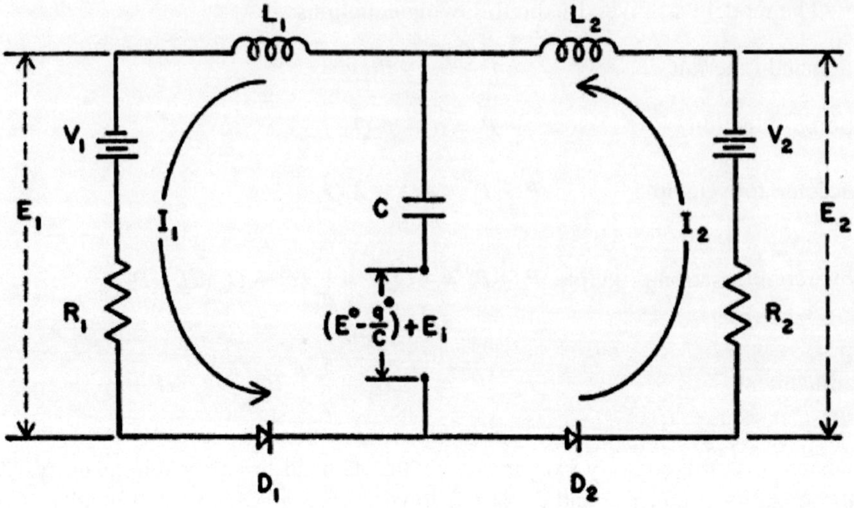

Fig. 7.3 Electrical equivalent circuit derived from the equations of the economic model. Source: Morehouse et al. (1950, 319).

This result was also presented in (Strotz et al. 1951), but this time to show electrical engineers what could be achieved with the Aeracom to investigate economic problems:

> 1. It may be desired to discover only general topological features of a model. Is the model stable or explosive? Does it give rise to a limit cycle, thus is it finite and periodic? How will variation in certain parameters alter the dynamic response of the model? To study the general topological features, it is unnecessary to have estimates of the actual values of the parameters, although the possible range of the parameters may be restricted by given inequalities.
> 2. For a given model it may desired to solve its system of simultaneous equations, and in so doing to employ those values for the parameters which have been obtained from independent statistical estimates.
> 3. For a given model [...] it may be desired to estimate the numerical values of the parameters [...], when the functions of time, with the exception of the stochastic variables, are assumed known over some historical period. (Strotz et al. 1951, 558)

Beside the model of inventory oscillations, Goodwin's (1951) nonlinear national income model and a simple two-equation national income model were investigated for these questions. A more extended analysis of Goodwin's model by Strotz, McAnulty and Naines with the aid of the Aeracom was published in *Econometrica* (1953).

To the 1951 paper (Strotz et al. 1951) a Discussion was added. Although called so, it was actually a brief note by Otto J.M. Smith[8] and R.M. Saunders, telling the

[8] Otto J.M Smith was Professor of Electrical Engineering at University of California, Berkeley and probably best known for the invention of what is now called the Smith predictor, a method of handling time delay, or deadtime, in feedback control systems.

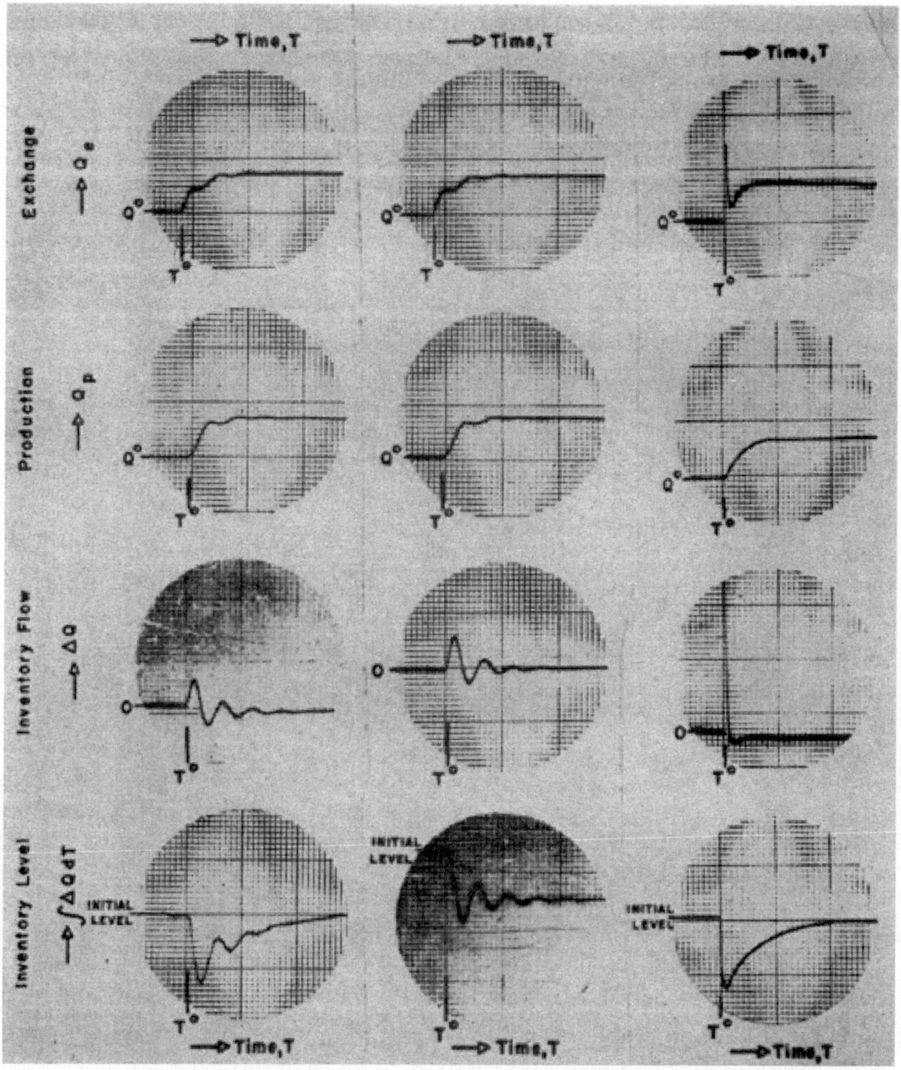

Fig. 7.4 Photographs of oscilloscope screens. Source: Morehouse et al. (1950, 321).

readers of the *AIEE Transactions* that they had applied analogue computing devices to study Kalecki's (1935) macrodynamic model. A 'relatively simple analogue' was used: 'a double-triode integrator, two tubes in the I-L subtractor, two phase inventors, and three adding tubes' (p. 563).

This simulation of Kalecki's macrodynamical system by an electrical circuit was published in a more extensive form in the journal *Electrical Engineering* in

1952, and now Smith's co-author was H.F. Erdley.[9] They gave the following motivation for their paper: 'An analogue is a way of thinking. It is a tool for visualization of a problem' (Smith and Erdley, 1952, 362).

Smith's (1953) 'Economic Analogs' was a survey article to inform engineers about the use of electric circuits to be used to solve economic problems:

> Although only a few specifically economic analogs have so far been built, and it is unlikely that many businesses can afford a general purpose unilateral analog computer, still, the analog techniques are very valuable to an economist because they give him a tool for thinking. (Smith 1953, 1518)

According to Smith the economic problems could be set up on any of the general purpose analogs like the BEAC, REAC, EASE or GEDA.[10]

To complete this survey of electronic analogs developed in the 1950s to solve economic problems, Stephen Enke's (1951) *Econometrica* paper should also be mentioned. This paper describes an electric circuit for determining prices and exports of a homogenous good in spatially distinct markets. Because no results were given it is not clear whether this circuit was ever actually built.

Last but not least there is, of course, Arnold Tustin's (1953) *The Mechanism of Economic Systems*, about 'the remarkable analogy that exists between economic systems and certain physical systems'. Although he referred only to a few examples, Tustin indicated some features of these analogues to explicate their potentialities for use in economics.

7.4.1 Cybernetics

The importance and relevance of using analogues in science was emphasised by the rise of new disciplines called 'cybernetics' and 'systems theory', and stimulated by successful experiences with operations research in World War II. In an early paper, even before the term 'cybernetics' was coined (by Norbert Wiener), Arturo Rosenblueth and Norbert Wiener (1945) wrote an article on 'the role of models in science'. They saw the intention and result of scientific inquiry to be obtaining understanding and control. But this could only be obtained by abstraction:

> Abstraction consists in replacing the part of the universe under consideration by a model of similar but simpler structure. Models, formal or intellectual on the one hand, or material on the other, are thus a central necessity of scientific procedure. (Rosenblueth and Wiener 1945, 316)

---

[9] The bibliography to this article mentions a M.S. Thesis by Erdley, *Business Cycle Stability from an Electrical Analog of an Economic System*. University of California, Berkeley, June 1950.

[10] BEAC: Boeing Electronic Analog Computer, Boeing Airplane Co; REAC: Reeves Electronic Analog Computer, Reeves Instrument Co; EASE: Electronic Analog Simulating Equipment, Beckman Instrument Co; GEDA: Goodyear Electronic Differential Analyzer, Goodyear Aircraft Corp. This information is from Berkeley and Wainwright (1956), published in a time when the number of computers in existence (about 170) still allowed such a survey.

They noted that not only models are abstractions, but 'all good experiments are good abstractions' (p. 316).

They distinguish between material and formal models:

> A material model is the representation of a complex system by a system which is assumed simpler and which is also assumed to have some properties similar to those selected for study in the original complex system. A formal model is a symbolic assertion in logical terms of an idealised relatively simple situation sharing the structural properties of the original factual system. (Rosenblueth and Wiener 1945, 317)

Material models are, according to the authors, useful in two ways. Firstly, 'they may assist the scientist in replacing a phenomenon in an unfamiliar field by one in a field in which he is more at home' (p. 317). Secondly, 'a material model may enable the carrying out of experiments under more favorable conditions than would be available in the original system' (p. 317).

W. Ross Ashby's (1956) *An Introduction to Cybernetics*, where the ideas and concepts of cybernetics were fully expounded, contains a section on isomorphic machines and a section on homomorphic machines. Two machines are isomorphic if one can be made identical to the other by simple relabelling. This relabelling can have various degrees of complexity, depending on what is relabelled: states or variables. If one of the two machines is simpler than the other, the simpler one is called a homomorphism of the more complex one. Though Ashby talks in terms of machines he also means to include mathematical systems and models. This is nicely illustrated by the following three isomorphic systems, see Fig. 7 5:

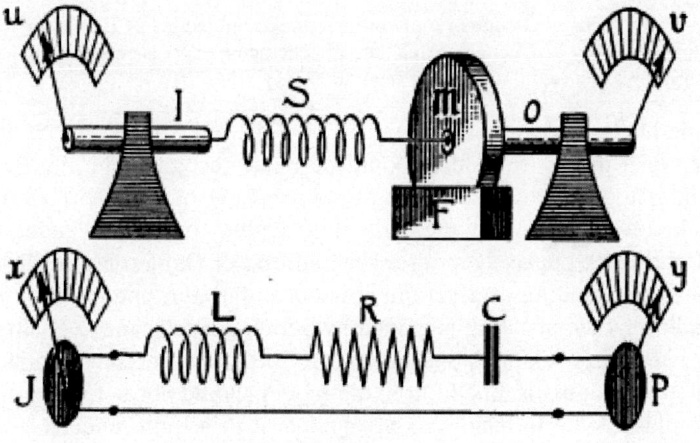

$$a\frac{d^2z}{dt^2}+b\frac{dz}{dt}+cz=w$$

Fig. 7.5 Three isomorphic systems. Source: Ashby (1956, 95–96).

Isomorphic systems are, according to Ashby, important because most systems have 'both difficult and easy patches in their properties'.

> When an experimenter comes to a difficult patch in the particular system he is investigating he may if an isomorphic form exists, find that the corresponding patch in the other form is much easier to understand or control or investigate. And experience has shown that the ability to change to an isomorphic form, though it does not give absolutely trustworthy evidence (for an isomorphism may hold only over a certain range), is nevertheless a most useful and practical help to the experimenter. In science it is used ubiquitously. (Ashby 1956, 97)

If interaction on a specific system is impossible, investigating an isomorphic, or homomorphic, system provides the same level of understanding.

7.5 Conclusions

With the international trend in the late 1980s for Central Banks to become more transparent, *De Nederlandsche Bank* aimed at a 'visual model' that would allow for 'more rapid and better insight' than 'a set of mathematical equations with explanatory notes'. MORKMON was built for the purpose of macroeconomic policy analysis, but because of its complexity it did not provide confidence to the Dutch citizens about how monetary policy choices were made at the Bank. The visual model was intended to be more intelligible than the econometric model of 164 equations. What they meant by 'more rapid and better insight' is close to De Regt's account of intelligibility. He suggests the following criterion for a theory to be intelligible:

> A scientific theory *T* (in one or more of its representations) is intelligible for scientists (in context *C*) if they can recognise qualitatively characteristic consequences of *T* without performing exact calculations. (De Regt 2009, 33)

While De Regt does not account for how the qualitatively characteristic consequences of a theory or model should be made recognizable, the Bank, from the beginning, aimed at a visualisation of the workings of MORKMON by developing an analogue model. With the aim of providing insight by using an analogue model, the Bank put itself in a longer tradition of Galileo-Maxwell-Boltzmann to gain an understanding of a yet unknown or unfamiliar phenomenon or system. In this tradition understanding is gained by building a substantive analogy, where the analogy transfers relationships representing experience and familiar concepts to the unfamiliar domain. In addition, the model should allow for interaction, that is, one should be able to handle or manipulate it to acquire understanding. So for a visual model to provide understanding it should represent familiar relationships and be manipulated.

The development of FYSIOEN shows that these elements were crucial; although it was an interactive model, and the images look like familiar hydraulic systems, its relations were not the familiar hydraulic principles but still the MORKMON relationships. So, although one could arrive at some level of understanding by playing with the visual model, this level of understanding is very much like the level of understanding one will arrive at with experimenting on the electronic analogues built in the 1950s. Notwithstanding the familiar look of the FYSIOEN images, the internal mechanisms of MORKMON remain invisible.

However, it might be that this kind of transparency is not needed to create trust in the Bank's policies. The current interactive program developed for educational purposes about the DNB's role and function in the Dutch economy is 'Scoren met beleid' (Score with policy). It is a game that can be played online.[11] Originally one was aiming at an update of FYSIOEN but 'gradually something more funny arose' (Albers 2006, 19, trans.). Like FYSIOEN all kinds of decisions can be made, but now the underlying model is completely blackboxed, only the decision results are given: A player in the role of advisor can choose among one of the presented scenarios and survey the macro-economic consequences of it. Subsequently comments by policymakers are displayed, which the player can use to compose an advice to the president of DNB. This advice is being evaluated by the program and 'paid-out' as an increase or decrease of salary. By playing it enough times, you learn how to increase your salary. Instead of aiming at understanding by 'ways of seeing and doing' one arrived with FYSIOEN and 'Scoren met beleid' at understanding only by 'ways of doing.'

References

Albers, J.: Scoor met beleid en sleep een virtueel contract binnen! DNB Magazine 3, 18–19 (2006)

Ashby, W.R.: An introduction to cybernetics. Chapman and Hall, London (1956)

Berkeley, E.C., Wainwright, L.: Computers their operation and applications. Reinhold Publishing Corporation, Chapman and Hall, New York, London (1956)

Bikker, J.A., Kramer, P.: Overleg te TH Delft op 22 januari. DNB Afdeling Wetenschappelijk onderzoek en econometrie, BI232 (1986)

Bissell, C.C.: Historical perspectives – The Moniac A Hydromechanical Analog Computer of the 1950s. IEEE Control Systems Magazine 27(1), 59–64 (2007)

Boltzmann, L.: On the methods of theoretical physics. In: McGuinness, B. (ed.) Theoretical Physics and Philosophical Problems, pp. 5–12. Reidel, Dordrecht (1892/1974)

Boltzmann, L.: On the development of the methods of theoretical physics. In: McGuinness, B. (ed.) Theoretical Physics and Philosophical Problems, pp. 77–100. Reidel, Dordrecht (1899a/1974)

Boltzmann, L.: On the fundamental principles and equations of mechanics. In: McGuinness, B. (ed.) Theoretical Physics and Philosophical Problems, pp. 10–128. Reidel, Dordrecht (1899b/1974)

Boltzmann, L.: Model. In: McGuinness, B. (ed.) Theoretical Physics and Philosophical Problems, pp. 213–220. Reidel, Dordrecht (1902a/1974)

Boltzmann, L.: On the principles of mechanics. In: McGuinness, B. (ed.) Theoretical Physics and Philosophical Problems, pp. 129–152. Reidel, Dordrecht (1902b/1974)

Boltzmann, L.: Anmerkungen. In: Boltzmann, L. (ed.) Faradays Kraftlinien von J.C. Maxwell, pp. 97–128. Engelmann, Leipzig (1912)

[11] See http://www.dnb.nl. Accessed 7 June 2011.

De Regt, H.W.: Ludwig Boltzmann's Bildtheorie and scientific understanding. Synthese 119(1/2), 113–134 (1999)

De Regt, H.W.: Understanding and scientific explanation. In: de Regt, H.W., Leonelli, S., Eigner, K. (eds.) Scientific Understanding, pp. 21–42. University of Pittsburgh Press, Pittsburgh (2009)

Dingemanse, L.J.: MORKMON-Visualisatie. Onderzoek naar de visualisatiemogelijkheden van het macro-ekonomisch model MORKMON. TU Delft, Afdeling Elektrotechniek nr. A 87.013 (1987)

Duhem, P.: The aim and structure of physical theory, translated by P.P. Wiener. Princeton University Press, Princeton (1954)

Enke, S.: Equilibrium among spatially separated markets: Solution by electric analogue. Econometrica 19(1), 40–47 (1951)

Fisher, I.: Mathematical investigations in the theory of value and prices. Yale University Press, New Haven (1892/1925)

Fisher, I.: The purchasing power of money, 2nd revised edn. Kelley, New York (1911/1963)

Fisher, I.: A three-dimensional representation of the factors of production and their remuneration, marginally and residually. Econometrica 7(4), 304–311 (1939)

Goodwin, R.M.: The nonlinear accelerator and the persistence of business cycles. Econometrica 19(1), 1–17 (1951)

Hertz, H.: The principles of mechanics presented in a new form. Dover, New York (1899/1956)

Kalecki, M.: A macrodynamic theory of business cycles. Econometrica 3(3), 327–344 (1935)

Keynes, J.M.: The general theory of employment, interest and money. Macmillan, London (1936)

Kramer, P.: FYSIOEN, een demonstratiemodel van de Nederlandse economie. DNB Afdeling Wetenschappelijk Onderzoek en Econometrie 8620 (1986)

Kramer, P., Bikker, J.A.: Stylering MORKMON. DNB Afdeling Wetenschappelijk Onderzoek en Econometrie, KR 151 (1986a)

Kramer, P., Bikker, J.A.: Stylering MORKMON (= MINIMORKMON). DNB Afdeling Wetenschappelijk Onderzoek en Econometrie, KR 159 (1986b)

Kramer, P., Coenen, R.L.: Werkprogramma ontwerp fysiek model MORKMON. DNB Afdeling Wetenschappelijk Onderzoek en Econometrie, KR 122 (1985a)

Kramer, P., Coenen, R.L.: Voortgang 'Fysiek model MORKMON'. DNB Afdeling Wetenschappelijk Onderzoek en Econometrie, KR 125 (1985b)

Kramer, P., van den Bosch, P.P.J., Mourik, T.J., Fase, M.M.G., van Nauta Lemke, H.R.: FYSIOEN. Macro-Economie in Computerbeelden. Kluwer, Deventer (1988)

Kramer, P., van den Bosch, P.P.J., Mourik, T.J., Fase, M.M.G., van Nauta Lemke, H.R.: FYSIOEN: Macroeconomics in computer graphics. Economic Modelling 7, 148–160 (1990)

Machamer, P.: Galileo's machines, his mathematics, and his experiments. In: Machamer, P. (ed.) The Cambridge Companion to Galileo, pp. 53–79. Cambridge University Press, Cambridge (1998)

Machamer, P., Woody, A.: A model of intelligibility in science: using Galileo's balance as a model for understanding the motion of bodies. Science and Education 3, 215–244 (1994)

Maxwell, J.C.: On Faraday's lines of force. In: Niven, W.D. (ed.) The Scientific Papers of James Clerk Maxwell, vol. I, pp. 155–229. Dover, New York (1855/1965)

Maxwell, J.C.: On the mathematical classification of physical quantities. In: Niven, W.D. (ed.) The Scientific Papers of James Clerk Maxwell, vol. II, pp. 257–266. Dover, New York (1871/1965)

Morehouse, N.F., Strotz, R.H., Horwitz, S.J.: An electro-analog method for investigating problems in economic dynamics: Inventory oscillations. Econometrica 18(4), 313–328 (1950)

Morgan, M.S., Boumans, M.: Secrets hidden by two-dimensionality: The economy as a hydraulic machine. In: de Chadarevian, S., Hopwood, N. (eds.) Models: The Third Dimension of Science, pp. 369–401. Stanford University Press, Stanford (2004)

Morrison, M., Morgan, M.S.: Models as mediating instruments. In: Morgan, M.S., Morrison, M. (eds.) Models as Mediators, pp. 10–37. Cambridge University Press, Cambridge (1999)

Mosselman, J.F.: Een interactief programma voor simulaties met MORKMON-deelmodellen. TU Delft, Faculteit der Elektrotechniek A 87.070 (1987)

Nagel, E.: The Structure of Science: Problems in the Logic of Scientific Explanation. Routledge and Kegan Paul, London (1961)

Oxford English Dictionary. Clarendon Press, Oxford (1933)

Phillips, A.W.: Mechanical models in economic dynamics. Economica 17(67), 283–305 (1950)

Reus, J.A.: Suggesties ter verbetering van de MORKMON-visualisatie. Vakgroep voor regeltechniek. TU Delft, Faculteit der Elektrotechniek T 87.042 (1987)

Rosenblueth, A., Wiener, N.: The role of models in science. Philosophy of Science 12(4), 316–321 (1945)

Small, J.S.: General-purpose electronic analog computing: 1945-1965. IEEE Annals of the History of Computing 15(2), 8–18 (1993)

Smith, O.J.M.: Economic analogs. Proceedings of the Institute of Radio Engineers 41(10), 1514–1519 (1953)

Smith, O.J.M., Erdley, H.F.: An electronic analogue for an economic system. Electrical Engineering 71, 362–366 (1952)

Smith, O.J.M., Saunders, R.M.: Discussion. Transactions of the American Institute of Electrical Engineers 70(1), 562–563 (1951)

Strotz, R.H., Calvert, J.F., Morehouse, N.F.: Analogue computing techniques applied to economics. Transactions of the American Institute of Electrical Engineers 70(1), 557–562 (1951)

Strotz, R.H., McAnulty, J.C., Naines, J.B.: Goodwin's nonlinear theory of the business cycle: An electric-analogue solution. Econometrica 21(3), 390–411 (1953)

Thomson, W.: Notes of lectures on molecular dynamics and the wave theory of light. In: Kargon, R., Achinstein, P. (eds.) Kelvin's Baltimore Lectures and Modern Theoretical Physics, pp. 7–255. MIT Press, Cambridge (1884/1987)

Tustin, A.: The Mechanism of Economic Systems: An Approach to the Problem of Economic Stabilisation from the Point of View of Control-System Engineering. William Heinemann, London (1953)

Chapter 8
The Inner World of Models and Its Epistemic Diversity: Infectious Disease and Climate Modelling

Gabriele Gramelsberger and Erika Mansnerus

Freie Universität Berlin and London School of Economics and Political Sciences

Abstract. Modelling and simulation techniques have various functions in scientific research. They may be used as measuring devices, tools, representations or experiments, or they may be regarded as 'artificial nature' that allows further investigation of a particular phenomenon. However, these functions vary according to the dominant field of research. Applied science, engineering and technology-driven applications develop and utilise modelling and simulation techniques in a unique way. For policy-driven research questions, the main interest extends beyond chains of plausible scientific inference. We will highlight this by characterising the unique aspect of modelling and simulation techniques, an epistemic diversity that is derived from the 'inner world of models', but which has implications for the applicability of the techniques.

8.1 Introduction

During the past decades computer-based simulations have increasingly gained ground in science. Nearly every discipline has turned into a computational one using numerical models to produce knowledge. The emerging computational departments of physics, chemistry or biology demonstrate the ongoing shift in science, with this development increasingly drawing the attention of philosophers of science. Whether simulation extends beyond the scientific methods of knowledge production (for example, theory, experiment, measurement) as a new and autonomous method, or whether it is a hybrid method 'somewhere intermediate between traditional theoretical [...] science and its empirical methods of experimentation and observation', is a question of particular interest to philosophers (Rohrlich 1991, 507). This epistemological question not only tries to specify how the simulations produce knowledge, it also aims to explore how simulation-based knowledge influences scientific practice and everyday life. The everyday impact of simulations has become tremendously important since simulations increasingly

C. Bissell and C. Dillon (Eds.): Ways of Thinking, Ways of Seeing, ACES 1, pp. 167–195.
springerlink.com

provide an evidence base for policy choices, as the cases from epidemiology and climate research will show.

This chapter aims to give some insights into the problem of using computer simulation-based knowledge for policy concerns. Major problems result from an insufficient knowledge about the inner mechanisms of simulation models, how they interact and how they influence the results. As philosophers are usually interested in a more general view, these inner mechanisms have not received enough attention, and simulation models remain as black boxes. Therefore, this chapter will open up two different kinds of models – infectious disease and climate models – in order to explore the mechanisms internal to the models, their influence on simulation results, and their relevance for policy concerns.

8.1.1 Philosophical Framework

Before going into the details of model-internal mechanisms, the framework of a current philosophy of simulation will be outlined. This framework is an open one, as there is no clear definition or theory as to what a simulation is. The heterogeneous scientific examples show that there is a class of different practices corresponding to the label 'simulation'. This is not surprising, as former studies on other scientific methods such as experiments (for example, Dear 1995) and models (for example, Morgan & Morrison 1999) have already shown. It is, therefore, very difficult to grasp the genuine epistemic properties of simulations, in particular as they contrast with models. The most basic property is described by Stephan Hartmann as imitating 'one process by another' – a natural process by a simulated one (Hartmann 1996, 77), to which he added more concretely, 'a simulation results when equations of the underlying dynamic model are solved (p. 83).[1] Paul Humphreys, too, tries to specify simulation more generally: 'A computer-simulation is any computer-implemented method for exploring the properties of mathematical models that are analytically intractable' (Humphreys 2004, 108). Both concepts refer to simulation as a way of handling specific – discretised, numerical – models that are too complex to be solved analytically. Both have models in mind that articulate theory within a century-old canon of differential equations. Climate models, for instance, belong in part to this class of simulation models, as they employ the Navier-Stokes equations for expressing the dynamics of the air masses' flow in the atmosphere.

However, identifying theory with differential equations, which can be re-articulated by discretised models, has led to the idea that simulation is a hybrid method, best described as 'experimenting with theories' (for example, Dowling 1999; Küppers and Lenhard 2005; Gramelsberger 2010). The problem here is that the terms 'theory' and 'theoretical' are often mixed up. In the philosophy of science 'theory' refers to a very general form of knowledge, whose role models

[1] Hartmann's account relates simulation to dynamic models, which are designed to imitate the time evolution of a real system. Here, the semantic view of theory structures is prolonged (Suppes 1960; van Fraassen 1980). In this view, a model is a mathematical structure used to represent the behaviour of a system, and this behaviour can now be displayed by performing simulation runs.

include Newton's mechanics, Euler's concept of fluid dynamics – expanded by Navier and Stokes – and Einstein's theory of relativity. Besides these general theories there are numerous empirical and phenomenological laws, assumptions rooted in local observations and measurements, and heuristics based on common sense. Simulation models incorporate all of these different knowledge sources, which determine the model-internal mechanisms, and this will be the main focus of this chapter.

Furthermore, the term 'theoretical' refers to every knowledge that is articulated semiotically. As simulation models are written down in computer languages they are necessarily theoretical, rather than experimental or empirical. Therefore, the concept of 'experimenting with theories' must be reformulated more accurately as 'experimenting with all kinds of scientific knowledge expressed semiotically'. And even more specifically as 'experimenting with all kind of scientific knowledge expressed semiotically in terms of machine algorithms'. We will see that not only the discretisation of models, but also machine algorithms contribute to the problems of inner-model mechanisms.

However, even the very term 'experimenting' has to be problematised. Usually experiments deal with real-world objects and phenomena and create semiotic results such as measurement data, which support theoretical explanations.[2] But experimenting with these semiotic results – data and theoretical explanations, which constitute an initialised simulation model – is something different. For instance, real-world experiments can fail, and nature can validate experimental designs and questions through its resistance. Experiments on scientific knowledge expressed algorithmically lack such instant validation. They can easily mislead the researcher by opening up a completely fictive realm of results, such that it is not easy to decide whether a result is valid or not. The whole debate on uncertainty follows from this basic problem.

Nevertheless, simulations open up what is, in principle, an infinite realm of results that can also be regarded positively as one of the genuine epistemic properties of simulations. This has led various authors to assign simulation to a 'third type of empirical extension' (Humphreys 2004, 5), asking whether 'instrumentally enhanced mathematics allow an investigation of new areas of mathematics to which our only mode of access is the instrument itself' (p. 5). While the first type is our bare perception, the second type is the typically scientific one: the instrument-mediated experience, therefore expanding the first type. Classic examples are the microscope and the telescope, which opened up new insights and worlds. However, the new third type is also instrument-mediated through its use of computers, which also open up new insights and worlds. But these new worlds are solely of a mathematical kind. In fact, the third type opens up the infinite space of numbers by computing trajectories into this space (Gramelsberger 2010). It is a new type of perception as it necessarily requires fast computers, particularly since 'when it comes to complex systems, we simply cannot bend our theories to our

[2] Therefore, simulation constitutes a new location of research, which is best addressed by the term 'dry lab'. Working in such a dry lab differs both from pure theoretical work and from work in a 'wet lab' (Merz 1999, 2002).

cognitive will—they will not yield results with any mechanical turn of a crank' (Winsberg 1999, 291).

Thus, simulation constitutes a second nature, leading authors such as Erika Mansnerus to claim that the classic scientific "metaphor of 'putting questions to Nature' could be upgraded [...] to the form of artificial nature formed by simulations" (Mattila 2006c, 91). Artificial here means, following Herbert Simon, 'man-made as opposed to natural' (Simon 1996, 5). Discoveries in this second nature are fuelled by the increase in computer power; to some extent this is comparable to the increase in resolution that granted access to ever smaller worlds through the microscope. Furthermore, they necessarily require experimental manipulation to explore the solution space of a simulation model – for alternative or future states – because the main benefit of 'putting questions to artificial nature' is that it opens up the explanatory capacities of these models for 'what would happen if' type of questions (Mansnerus 2011). Real-world experiments can explore such questions only in part, but they are at a complete loss when this type of question is projected from the future onto our present as climate models do. These predictive questions constitute another genuine epistemic property of simulation by scanning the space of states of a system under varying initial and boundary conditions into the future (Woodward 2003; Mattila 2006a, b).[3]

8.1.2 Model-Internal Mechanisms

A large body of literature in philosophy deals with the ontological status of simulation, as already outlined, aiming to clearly differentiate between empirical and theoretical methods respectively. The epistemological constitution of simulation-based results has been explored too, as well as questions of validation and evaluation. Thereby the simulation models, mostly borrowed from physics, are seen as a closed body of theory from which statements can be derived by creating a 'complex chain of inferences that serve to transform theoretical structures into specific concrete knowledge [...] that has its own unique epistemology', which is downward, autonomous and motley (Winsberg 1999, 275).[4] However, as simulations are usually large bodies of computable texts describing complex phenomena, this view of models from the outside is not sufficient. Simulations tell a much more complex story (Morgan 2001). But how are the stories told by modellers?

---

[3] The novelty of this kind of question is relative to the field that applies the simulation techniques. In epidemiology, and especially in the modelling research conducted at the National Institute for Health and Welfare in Finland, simulation models are a fairly recent development, whereas in economics Meade already used these manipulative questions back in the 1930s, as shown in Morgan (2001).

[4] With 'downward epistemology' he means that in physics, which is the subject of his case, simulation models are based on higher-level theories about the physical world and used, at least in part, to explain phenomena in terms of mechanisms. 'Autonomous' refers to the fact that simulation models are constructed in order to extend theories into new domains of application. With 'motley' epistemology Winsberg also acknowledges that model building draws on a variety of sources, which include not only theory, physical insight, extensive approximations, and idealisations (Winsberg 2001; 2009).

In order to answer these questions, the role of the various components involved in simulation models must be investigated. This can be done by taking a closer look at the practices of modelling and coding. Therefore, active investigation of and access to the models' code are necessarily required, as this is not a typical domain for philosophers. Nevertheless, Marcel Boumans has studied the practice of modelling small business-cycle models, and he suggests that models gain their justification through the diversity of 'built-in ingredients' like data, theoretical assumptions, policy views, metaphors, and mathematical formulae. These ingredients, according to him, are 'baked into a model without a recipe' (Boumans 1999, 67), so that expertise is required to apply them to the outer world, for instance by addressing policy-driven questions in the model. The listed ingredients denote some of the components of a simulation model, but not all. It turns out, when a closer look is taken at the models' code, that simulation models telling a complex story employ epistemically diverse and heterogeneous knowledge sources. Theories are just one of these; the others include assumptions, heuristics, and estimations.

8.1.3 The Case Studies

The aim of the following investigation is to open up the black box of simulation models and to give some insights into the 'epistemic diversity' of this inner world of models. This is possible only for very concrete examples. The case studies are taken from two very different fields: epidemiology and climate research. Their conceptual frameworks could not differ more. Yet, both fields share the similar attention outside the scientific world as they are important tools for providing an evidence base for policy options. Therefore the main research question is: how can the heterogeneous and diverse inner-world of simulation models create reliable results that support policy decisions? This question is usually answered by focusing on the topic of uncertainty (for example, for climate research see Gramelsberger and Feichter 2011), but the following investigation explores a different perspective by focusing on the practices of modelling and coding. The conceptual framework, therefore, follows a top-down analysis from the entire model to the epistemically diverse ingredients that constitute the model (data, theoretical assumptions, policy views, metaphors, and so on) and finally to the statements of the code that semiotically articulate the ingredients on a machine algorithm level. This conceptual framework is rooted in the scientific practice of modelling, coding and using simulation models as well as in the different functions which a simulation model has to fulfil: it has to be applicable for 'putting questions to artificial nature', such as policy concerns; it has to be conceivable in order to create a specific artificial nature or experimental setting; and it has to be computable, as the 'mechanical turn of a crank' is indispensable. While the statements (computable version) and the ingredients (conceptual version) constitute the inner world of a model, the entire model – as applied for simulation runs – must encompass the outer world, delivering results and answers (see Table 8.1). In other words, on the level of using models for policy-making, epistemic diversity operates when models are applicable to policy-driven questions. This is what we refer to as the outer world of models. The inner world is then composed of ingredients that are conceivable for the modellers, and statements that are translations of ingredients into code and hence computable.

Table 8.1 The inner and outer worlds of simulation models.

Relations	Levels	Functions
Outer world	model	applicable
Inner world	ingredients	conceivable
Inner world	statements	computable

8.2 The Inner World of Infectious Disease Models (Ingredients)

Computer-based modelling and simulation techniques are finding increasing use in infectious disease epidemiology as a source of evidence for optimising vaccination schedules or planning mitigation strategies for pandemic outbreaks. The most recent example comes from pandemic-preparedness simulations that were used to study the effectiveness of mitigation strategies, such as travel restrictions or quarantine, although models are also used effectively to re-assess vaccination strategies. Modelling allows epidemiological questions to be explored that might be difficult to study because of financial and ethical restrictions. This section will analyse how modelling methods developed as a part of the research agenda at the National Institute for Health and Welfare, Helsinki, Finland. Through this analysis of the micro-practices of modelling we will learn what infectious disease models are, how they are built and what historical developments in epidemiology can be seen as a stimulus for developing mathematical techniques in infectious disease epidemiology. These aspects will show how epistemic diversity is manifested in infectious disease models.

8.2.1 Towards Mathematical Representations of Infections

Mathematical representations of infections arose from the interest in quantifying observations of human diseases. Daryl Daley and Joe Gani trace the history of these studies back to John Graunt's 1662 work, 'Natural and Political Observations made upon the Bills of Mortality'. Graunt's work created a system for warning of the spread of the bubonic plague. Following that, a century later, Daniel Bernoulli demonstrated how variolation[5] could reduce death rates; and in the 1840s, William Farr studied the progress of epidemics and characterised data from smallpox deaths mathematically. These were limited approaches, since the understanding of the mechanisms that spread the disease had yet to be discovered (Daley and Gani 1999).

[5] Inoculation of smallpox virus from a diseased person to a healthy person.

Once the causes of disease began to be understood, providing for a mindset that allowed transmission dynamics to be expressed in terms of spreading *germs*, the mathematical theory of epidemics began to emerge. Pioneering work by William Hamer elaborated the mass-action principle in a deterministic model of measles outbreaks. His work, 'Epidemic Disease in England – the Evidence of Variability and of Persistency Type', studies the observations from a London measles wave, for which he is able to identify factors affecting population-level transmission: 'Alterations of its age constitution, varying customs, and social conditions' (Hamer 1906).

The periodic alteration of the infection was already acknowledged by Herbert Soper as being of great interest to human experience (Soper 1929). One could say that it was this periodicity that led the pioneers of mathematical epidemiology to observe the patterns of transmission and express them in mathematical terms, which can be called early models. For example, categories of susceptibles, infected, and immunes within a population were identified in order to follow the development of an outbreak (Kermack & McKendrick 1927). What were then the main contributions of these mathematical expressions?

For Hamer (1906), it was not enough to explain singular causes of diseases but to find the explanation why transmission patterns circulate between populations. In order to improve the understanding of disease transmission, this population was represented as sub-groups or *compartments* to show how different groups of individuals become infected.

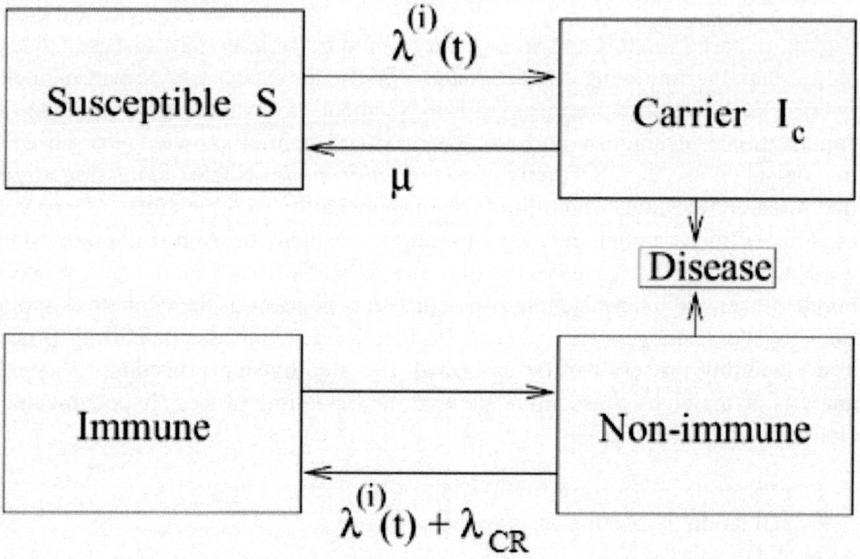

Fig. 8.1 Illustration of compartments that divide the population into subgroups of susceptibles, carriers, immune and non-immune. Transitions between these sub-groups can be expressed as rates of infection in relation to time. Source: Auranen, unpublished manuscript.

What, then, are compartmental models? In simplistic terms, they are models that consider groups of individuals in a population as compartments, which are separate sets that can be described in relational terms (for example, transition from susceptible to immune as a function of time). By naming these compartments according to the state of infectiousness, one is able to divide the population into smaller units and express (with differential equations) the transitions between these units in relation to time, as in Fig. 8.1.

Diagrammatically, movement between compartments is usually expressed by arrows indicating changes in the daily rates (Hurst & Murphy 1996). The usual compartments are susceptible, infectious and immune. These basic categories can be extended to infected (but not yet infectious), removed, or deceased; yet another category can be added to describe those who are protected by maternal antibodies (immune, but not due to disease). In addition to these categories, compartmental models describe the interrelations: The transitions from susceptible to infected to immune. The rates of these transitions can then be diverted by interventions: Adding vaccinations to the model increases the number of the immune in the population, for example. Compartmental models can be seen as predecessors of dynamic transmission models that became part of epidemiological research in the late 1970s. These are now dominant in infectious disease modelling and represent the group of models that were built in Helsinki 1995–2001.

8.2.2 The Inner World of Infectious Disease Models

Computer-based models and simulation techniques in infectious disease epidemiology share the following characteristics. Firstly, they have a three-part elementary structure, which comprises a data element, a mathematical method and computational techniques, and an element of substantial knowledge or an epidemiological component. Secondly, they are *tailor-made*, usually addressing a specified research question, which limits their applicability to some extent. Thirdly, the majority of these models rely on currently available data. And it is precisely this need to re-use and re-analyse the data that is part of what motivates the model-building exercise. Fourthly, micro-practices independent of the context of application, say the pathogen studied, can be identified within the modelling process. The modelling process can be described as a step-by-step procedure. A detailed analysis of the eight consecutive steps in the modelling process is documented in Habbema et al. (1996).

1. Identification of questions to be addressed.
2. Investigation of existing knowledge.
3. Model design.
4. Model quantification.
5. Model validation.

6. Prediction and optimization.
7. Decision making (on the basis of the model output).
8. Transfer of simulation program (to be applied to another infection).

Habbema et al. provide one formal model of infectious disease modelling. It highlights the different steps or stages upon which most accounts of modelling agree, but it can be misleadingly reductionist if it is read as a linear procedure. This step-by-step description is a useful illustration of the various stages that take place in the modelling process. We argue that the main importance is the question. This follows from the idea of *tailoring* a model to address particular interests. The investigation of existing knowledge is a process in which existing literature, laboratory results, experiences of existing models and data from surveillance progammes are integrated as a part of model assumptions.

Mary Morgan (2001) aligns model-building with steps similar to those mentioned by Habbema et al. (1996), although her focus is on economic models. The main difference is that in her account the model is first built to represent the world, then subjected to questions and manipulation in order to answer the questions; after that the model output is related to real-world phenomena.

In contrast, model design in infectious disease epidemiology follows the existing understanding of how the phenomenon of interest behaves and is often represented through a compartmental structure. Model quantification is the process of estimating the optimal parameter values, and setting the algorithms to run the simulations. In the account by Habbema et al. model validation means checking the model against data from a control program. However, model validation is a broad and contested notion and some authors address it in terms of the reliability of models (Oreskes et al. 1994; Boumans 2004). By transferring the simulation program, Habbema et al. refer to the generalizability of the computer program for other infectious diseases.[6] This step-by-step characterisation of the micro-practices of modelling highlight that modelling can be associated with *tinkering*, which builds upon and checks back with previous steps throughout the process. Knowledge is produced while tinkering. Importantly, these models are not only scientific exercises to develop better computational algorithms, they are built first and foremost to explain, understand and predict the infectious disease phenomena of interest. The major application of this group of models (including simulations as well) is, for example, to design reliable and cost-effective vaccination strategies or to predict the course of an outbreak.

From these general observations of what disease transmission models are we will move on to the detailed analysis of a family of models related to *Haemophilus influenzae* type b.[7] What, then, are these models? In general terms, they are

[6] Model quantification is known in climate models as model parameterisation.

[7] A total of 15 models were built during the project. Most of these were of Hib-related research questions, but methods of modelling other bacterial agents or chronic disease were also developed. For the sake of clarity, the focus in this study is only on Hib models.

probabilistic transmission models of the bacterial pathogen Haemophilus influenzae type b (Hib), which can cause severe or life-threatening diseases such as meningitis, epiglottitis, arthritis, pneumonia and septicaemia, especially among infants and children. However, these severe conditions are rare due to the wide coverage of Hib vaccinations, which began in the mid-1980s. The aim in the modelling was to enhance the understanding of the dynamics of Haemophilus infection and to assess its persistence in a population (Auranen et al. 1996). A further objective was to 'develop methods for the analysis of Hib infection and the effect of different intervention strategies', formulated in a research plan in 1994 during a long-standing Finnish research project called INFEMAT.

Researchers from the Department of Vaccines at the National Institute for Health and Welfare (KTL), the Biometry research group at the University of Helsinki, the Rolf Nevanlinna Institute (RNI)[8], and the Multimedia Laboratory of Helsinki University of Technology (HUT) all participated in the project. The original initiative for the research came from the National Public Health Institute, Finland, which was able to find motivated researchers among those studying modelling, such as mathematicians working on electromagnetism, at the Rolf Nevanlinna Institute. The Multimedia Laboratory (HUT) represented expertise in simulation techniques applied in their studies of modelling artificial life.

This analysis contrasts Boumans's (1999) core idea, which is to understand models as *moulded* entities constituted of heterogeneous *ingredients* built in a manner analogous to 'baking a cake without a recipe' (Boumans 1999), with the idea of *tailoring* a model. Whereas Boumans suggests moulding as a practice to guide modelling, we consider model building to be an intentional practice, which is guided by *research questions*. These questions are formulated as assumptions that guide the process of choosing and combining the set of ingredients in the model. The questions as guiding forces in the modelling process highlight the idea of *tailoring*: models are built for a particular purpose or use. In discussing these translations from the questions to the ingredients in detail, this section introduces the life-span of the family of Hib models, and then analyses two models in detail, the 'Goodnight Kiss' model and the 'Simulation' model, in terms of their ingredients and how the questions relate to them. The life-span of the modelling could be divided up into four phases based on the principle goal of each phase. A summary of the phases, the models built in each phase and their ingredients is presented below in Table 8.2.

---

[8] The RNI became part of the Department of Mathematics and Statistics on 1 January 2004.

Table 8.2 Phases of modelling, the models built and their ingredients. Source: Mattila (2006).

Phase of Modelling	Models built	Ingredients
Phase I The Construction of the Goodnight Kiss Model (1993-1994) The INFEMAT project began. The researchers started to construct the first transmission model.	1. The first transmission model, later called the Goodnight Kiss Model.	The Bayesian and frequentist approaches to statistical modelling, mainly contributing to the development of modelling methodology in mathematics. Background assumptions about various diseases. Multiple datasets collected by KTL.
Phase II Building the Family of Models (1995-1997). Emphasis on the development of modelling methods.	Published version of the Goodnight Kiss Model (1995) (see Sect. 8.2.2.1) 2. Models of Hib antibodies.	A stochastic model. Computer-intensive methods and algorithms. The SIS[9] model as a basic epidemiological model. Two datasets: collected by KTL and by Prof. Barbour's research group in the U.K.
Phase III Interacting and Overlapping Models (1998-2000) The focus on epidemiological studies using the previously published models	3. The Hib carriage model 4. The Bayesian model for predicting the duration of immunity to Hib. 5. The age-specific incidence of Hib. 6. The dynamics of natural immunity.	Epidemiological assumptions on herd immunity and vaccination efficiency. Databases from KTL and the collaborative partners.
Phase IV The Emergence of the Simulation model (2001-2004). In the final phase, the simulation model and a related computer program were developed.	7. The model of cross-reactives in Hib infection. 8. The Hib immunity and vaccination model. 9. The prevention of invasive Hib infections by vaccination. 10. The population-structured Hib transmission model (Simulation model, see Sect. 8.2.2.2).	The previous ingredients: epidemiological assumptions and data, mathematical methods. Additional ingredient: simulation methods and techniques.

8.2.2.1 The Goodnight Kiss Model as an Example of a Simple Transmission Model

The Goodnight Kiss model (GNKM) provides an example of a simple transmission model and thus allows us to analyse its composition. What was the starting

[9] The susceptible-immune-susceptible (SIS) model describes the transitions between the two epidemiological states of an individual.

point? The questions set out in the original research plan for the project emphasised the importance of studying the different aspects of infection and its control in interventions:

- Does vaccination alter the age distribution of Hib diseases and its incidence?
- Does vaccination alter the spectrum of Hib diseases?
- Does natural immunity vanish from the general population, which would indicate the need for revaccination?
- How high must vaccination coverage be in order to prevent Hib diseases in the population?

The Goodnight Kiss model functioned as a basis for estimating family and community transmission rates simultaneously. It was built to promote understanding of the dynamics of Hib infection, and to evaluate the ability of the infection to persist in a population. The population under scrutiny was limited to a family, which has a simple structure for following transmission. The underlying assumption was that the spread of Hib carriage took place in goodnight kisses among members of families with small children, who were in the 'risk group' of contracting the Hib infection.[10] Thus, in order to address the more sophisticated research problems formulated in the plan, the modellers started with this more simple transmission model. Three groups of ingredients can be identified: statistical methods and solutions (including computational techniques), epidemiological mechanisms, and data.

The statistical ingredient in the GNKM applied the Markov process to model Hib carriage in a family. The Markov process is a stochastic process that can be defined in terms of random variables. In this case, the modellers were able to express the likelihood even though possible transitions (between the different states of carriage) were not recorded in the data. In fact this description of the statistical ingredients implies their intertwinement with the epidemiological ingredients. The epidemiological mechanisms, and more broadly the phenomenon-bound ingredients, were the background assumptions about the behaviour and transmission of Hib. The researchers introduced a simple compartmental pattern as an epidemiological mechanism in the following way: 'We consider an SIS-type epidemic model, where the infection states are non-carrier (S), that is who is susceptible to becoming a carrier of the bacteria, and infectious carrier (C), that is who is able to spread the bacteria' (Auranen et al. 1996, 2238). Therefore, with this SIS-type model the transitions between the individual as a non-carrier, an asymptomatic carrier or one with the rare invasive infection, were differentiated and the transmission rates calculated.

The Goodnight Kiss model incorporated two sets of data. The first set was collected as part of a risk-factor analysis of invasive Hib disease in Finland during 1985-1986, just before the Hib vaccination programme was launched, and the second set was collected in the United Kingdom during 1991-1992. The data were collected from infants and family members when the infants were six, nine

[10] According to Auranen et al. (1996), 95 per cent of all invasive diseases occur among children under 5 years of age.

and twelve months of age. Composed of these three sets of ingredients, the Goodnight Kiss model offered an example of individual-based modelling in a small population.

8.2.2.2 The Simulation Model: The Emergence of a Multi-layered, Computer-Based Model

The individual-based stochastic simulation model, abbreviated as the Simulation model, which was built during the final phase, represents a more complex model than the GNKM. Characteristic of the model is its three-part structure, which consists of a population model, a transmission model and an immunity model. We will first describe the three sub-models, and then identify the ingredients.

- The population model is a demographic model that produced a population comparable to the Finnish population based on data collected by Statistics Finland. It depicted the size of the contact sites and the corresponding age groups. Participation in the contact sites as a member of a day-care group or a school class was age-related. The corresponding family size and structure were also built into the demographic model.
- The transmission model naturally described the transmission in the contact sites, as the Goodnight Kiss model had already done, and it also brought the simulation into play: it monitored all contacts and counted the proportion of carriers among them. A small, constant background intensity of exposure from the whole population was incorporated into the simulation.
- The immunity model described immunity against invasive Hib diseases, which was produced in response to Hib and cross-reactive bacteria[11] (Mäkelä et al. 2003). It also examined the changes in inherited immunity and monitored its waning over time. Because immunity is 'controlled' and 'boosted' in individuals by vaccinations, the vaccination model is a part of the immunity model. Hib vaccination reduces the incidence of Hib carriage and, even when it does not succeed, it is likely to prevent the progression from carriage to invasive disease. The vaccination model concerned the Finnish vaccination strategies of the 1980s, and was also sensitive to different types of vaccines.[12]

Thus, the Simulation model was composed of three sub-models, but what were its ingredients? This question is crucial, because the ingredients from the previously built models were moulded and reformulated as parts of these three sub-models. This provides an opportunity to see how a family of models represents a continuity that carries the ingredients from the early phases to the final one, from the simple, singular models to this complex, multi-layered final one. Each of these sub-models shares the statistical, epidemiological and data ingredients. What had been learnt about the population dynamics, the transmission and spread of Hib bacteria, the vaccination effects and their influence on herd immunity were to

[11] Cross-reactive bacteria, or CR, are antibodies that also react with some antigens other than the ones they were produced in response to (Mäkelä et al. 2003).

[12] These being conjugate and polysaccharide vaccines.

some extent encapsulated as ingredients. The datasets were affected by the problem of missing data, due to the fact that the data were originally collected for different purposes.

The problem is that the stocks of previously collected data do not document all the traits of the phenomenon that might have been useful in building the current model. The solutions benefited from Bayesian data augmentation. As Auranen explains: 'Unobserved events are not only incorporated in the model specification stage but often they are explicitly retained in the model as unknown observables' (Auranen 1999).

The core question in Bayesian inference concerns how to interpret probability. Bayesians consider probability a personal, subjective opinion of how likely the happening would be. This personal view is updated, changed as evidence, and through data, accumulates.[13] Furthermore, Bayesian hierarchical modelling was described as the methodological approach applied in the models. 'Hierarchical modelling enhances data augmentation where underlying infectious processes are identified explicitly' (Auranen 1999, 9). The built-in hierarchy strengthened the reliability of the models, and facilitated description of how one person infects another. How were these ingredients represented in the family of models?

8.2.3 Summary – From Ingredients to Models

This family of models could be summarised as follows. Let us consider the Simulation model as an endpoint. Each of its sub-models contained ingredients from the models built earlier. The demographic sub-model was expanded from the simple family structure presented in the Goodnight Kiss model. In a similar way, the transmission patterns studied primarily in a closed population (in the GNKM) were extended to cover the complicated contact-site structure and corresponding age-dependency of day-care groups and school classes. In addition, the question of immunity to Hib was studied in a model (see Table 8.2, no. 6), called the 'Dynamics of natural immunity' in the Simulation model. The key findings of the Dynamics model were scaled up to mimic the development of an individual's immunity combined with the vaccination effects. In fact, the details of the dynamics of immunity, the vaccination effects and the role of cross-reactive bacteria were explored in the whole set of built models (see Table 8.2, nos. 2-9), all of which shared the ingredients of statistical methods, epidemiological assumptions, and patterns and data. The main difference between these and the Simulation model lay in the additional ingredient, namely that the Simulation model was implemented in a computer program. This program was used for generating data to project the possible future developments of the bacterial behaviour and its encounters with human beings.

Hence, the family of models originated from questions that were translated into ingredients, and the ingredients then functioned as the building blocks for the set of models. These models provided a starting point for building the Simulation

[13] As the modellers expressed it, 'Let the data speak for themselves', which took on another meaning as a slogan in the course of my study (Mattila 2006c).

model. Although the ingredients were scattered throughout the models, they represented a miniature of the question to be studied in the Simulation model. Thus, the notion of a family of models emphasises the mutual continuity of models throughout the long-standing research project. This continuity could be conceptualised in terms of the necessary and interdependent functions of the ingredients. In countering Boumans' (1999) argument concerning model ingredients and his metaphor of modelling as 'baking a cake without recipe', we maintain that the ingredients are interdependent and necessary, and that they are not arbitrarily built into the models, but carefully and intentionally chosen in order to answer the question posed.

The ingredients of the infectious-disease models were statistical methods, epidemiological mechanisms and background assumptions, and epidemiological data. These ingredients were not ready-made, but were developed in relation to the question(s) addressed in the model under construction. As we learnt from the Goodnight Kiss model, the transmission patterns were studied within a closed population, a family. This required the development of the Markov process in order to model the transmission rate. The epidemiological assumptions guided our understanding of the transmission mechanism (for example, that it required intimate contact between humans to be transmitted), and the datasets required resolution of the problem of missing data due to the fact that they were recycled, originally having been collected for different purposes. None of these ingredients alone was capable of comprising a model. Each model is an intelligent sum of its ingredients, which means that the ingredients are moulded, combined and integrated in a purposeful manner. They are thus necessary, and simultaneously they are interdependent, which gives a detailed explanation of their interaction with the questions asked in the model.

8.3 The Inner World of Climate Models – Statements

Ingredients, as outlined by Boumans (1999), refer to data, equations, theoretical assumptions and so forth, and thus unveil the specific epistemic diversity of business-cycle models, which he has analysed. As the case of infectious disease models has shown, their specific epistemic diversity is different from those of Boumans' models. Furthermore, it has been demonstrated that the ingredients are inherited from one model to the other in case of the multi-layered simulation model. The case of climate models will also present a specific composition of ingredients, but it will go further and analyse how the ingredients are translated into computable statements. This closer focus on the inner world of simulation models will show that the practice of simulation modelling is mainly a practice of concretisation, as the computation of simulation models requires concrete statements and parameter values. However, concretisation is needed, in turn, to build adequate simulation models, which tell a concrete and reliable story about real circumstances. Of course, the reliability of simulation models, as the final discussion will emphasize, is achieved differently in both fields, although reliability is the same basic precondition for both fields when these models are used to address policy concerns.

8.3.1 Towards Mathematical Representations of Climate

Until the beginning of the 20th century meteorology was an empirical and descriptive science, which lacked a sound theoretical foundation. The atmosphere as a fluid phenomenon of travelling air masses was far too complex to be described theoretically, although in the 18th century physics had already developed a theory of fluid dynamics based on a set of differential equations. But these equations could be applied only to very simple and idealized cases, such as the laminar flow of a homogeneous fluid in a straight tube without obstacles and friction. However, the atmospheric situation was an entirely different one, as friction and turbulences create the various weather phenomena such as the convection of warm and humid air masses. Therefore, it took a long time before meteorology could be based on a sound theoretical foundation. Yet a theoretical foundation is the indispensible precondition for modelling and, finally, for using these models for predicting future developments.

This theoretical foundation was outlined for the first time by the Norwegian physicist Vilhelm Bjerknes in 1904, when he conceived the atmosphere as a giant air circulation and heat exchange machine (Bjerknes 1904). His physical and mechanical view defined the atmosphere as the relation between the velocity of air masses, their density and pressure (three hydrodynamic equations of motion); the continuity of mass during motion (the continuity equation); the relation between density, air pressure, temperature and humidity of any air mass (equation of state for the atmosphere); and how the energy and entropy of any air mass change in a change of state (the two fundamental theorems in the mechanical theory of heat).

Thus, the state of the atmosphere is determined by seven meteorological variables: velocity (in three directions), and the density, pressure, temperature, and humidity of the air for any point at any particular time. Bjerknes' model of the atmosphere – based on thermo- and hydrodynamical theory and articulated by a set of seven differential equations – marked the turning point from meteorology to the physics of the atmosphere. His model is the core of every weather and climate model even today, and is the basis of all scientific weather predictions and climate projections. It has transferred the observed atmospheric phenomena into computable ones.

However, before Bjerknes' model can be used to compute the seven meteorological variables, two things are required. Firstly, the mathematical model, which is a continuous-deterministic description of any state of the atmosphere, has to be transformed into a discrete and computable version for specific states of the atmosphere. Therefore, the differential equations have to be replaced by difference equations (discretisation). In order to facilitate the formulation of differences the model domain has to be subdivided into a finite number of grid boxes; continuous variables such as temperature, density, and humidity have to be converted into discretised differences. After this, the discretised model can be solved for grid points, such that the results represent grid-box averages. These results have to be computed for each time interval, thus unveiling the model's behaviour over time.

Secondly, automatic computing machines are required to solve the discretised equations for as many grid points as possible in order to obtain meaningful results.

While the very first meteorological model, computed on ENIAC in 1950 by Jule Charney and John von Neumann, had to start with a 15 × 18 grid for one layer at 500 millibars covering the area of North America (Charney et al. 1950), the increase in spatial and temporal resolution has driven numerical weather forecasting and climate projection, and still does. Higher resolution enables a better 'image' of the computed atmospheric processes, and this increase in resolution is a direct result of the tremendous improvement in computing speed. Today's weather models usually consist of a resolution as fine as six kilometres, while climate models vary between 500 and 60 kilometres depending on the time period they compute, which can be some decades or centuries.

These conditions define the set of ingredients 'baked in' a weather or climate model. In contrast to Boumans' analysis, climate modellers have an advanced practice or 'recipe' for baking these ingredients into their models. A weather or climate model consists of two parts that are clearly distinguished by the resolution of a computed model: the adiabatic part (resolved processes) and the non-adiabatic part (unresolved processes). The ingredients of the first, adiabatic part are thermo- and hydrodynamical theory, which constitute the general form of knowledge on atmospheric processes employed in the model. The ingredients of the second, non-adiabatic part cover every process that takes place on a scale smaller than the model's resolution (subscale processes). This is necessary because the scale of meteorological effects ranges between centimetres (micro turbulences) and several thousand kilometres (planetary waves).

The chosen resolution of a model introduces an artificial cut into these two domains. For instance, the climate models of the first Assessment Report of the Intergovernmental Panel on Climate Change (IPCC) used a 500 km horizontal resolution, while the current 2007 report has improved resolution to a 110 km grid (IPCC 2007, 113). Despite this increase in resolution, explicit mathematical terms are needed to express the influence of unresolved processes on the resolved ones; these terms are called 'subscale parameterisations'.

The knowledge employed in these subscale processes is based not on thermo- and hydrodynamical theory, but on other sources of scientific knowledge such as information derived from measurement campaigns, laboratory experiments, higher-resolved simulations, or simple assumptions.[14] Thereby, the main inconvenience is that the subscale parameterisations bring with them many problems, as they introduce to meteorological models an epistemically diverse body of knowledge based on all sorts of scientific knowledge, which is afflicted with large uncertainties. It is these uncertainties that largely diminish the reliability of climate models.

The next section will take a closer view of this epistemically diverse body of knowledge by analysing the important parametrisation of clouds in climate models. Clouds have a strong effect on climate as they interact with atmospheric processes, but the global resolution of a climate model is still too coarse to

[14] The adiabatic part refers to the notion of 'theory' in the narrow sense (see Sect. 0.1 of this chapter), while the non-adiabatic part refers to the notion of 'theory' in the sense of 'theoretical'. That is, a semiotic representation of an epistemically diverse collection of scientific knowledge based on local measurements, assumptions, heuristics, and so on.

grasp their influence. Or put it differently, clouds are too small to be resolved in a climate model. Therefore, the effects of clouds on the resolved meteorological processes have to be expressed explicitly within the terms of the subscale parameterisation.

8.3.2 The Inner World of Climate Models

A brief overview of the history of cloud parameterisation shows that, depending on computer resources and the available empirical knowledge on cloud processes, the conceived cloud schemes refer to different modelling strategies such as prescribing, diagnosing or prognosticating. For instance, in early models the distribution of clouds was prescribed (Manabe et al. 1965). Later, cloud cover was diagnosed from large-scale atmospheric properties, such as relative humidity or vertical velocity computed by the adiabatic part (Smagorinsky 1963). As Michael Tiedtke, responsible for the cloud parameterisation of the weather forecast model of the European Centre for Medium-Range Weather Forecasts (ECMWF), pointed out, diagnostic schemes are 'successful in reproducing some of the gross features of global cloudiness, but they lack a sound physical basis and in particular do not represent the interaction between clouds and the hydrological cycle of the model. However, because diagnostic schemes are simple and yet rather successful, they are widely used in large-scale models' (Tiedtke 1993, 3040). Therefore, since the 1970s cloud parameterisations have been increasingly based on a sound physical basis, with the ultimate aim becoming a prognostic scheme for important cloud variables (for example, cloud cover, cloud water/ice content), whose time evolution is fully determined by the advective processes computed in the adiabatic part of the climate model.[15] However, an accurate prognostic scheme depends on a realistic representation of advective transports of cloud variables within the adiabatic part, but – based on the presence of discontinuities – the numerical schemes for advective transport can cause large truncation errors, leading to unrealistic estimates of cloud water content and, in consequence, to an unrealistic prognosis of cloud variables.

Not only the available empirical and physical knowledge constrains modelling strategies, problems due to computation and discretisation also determine the practices of modelling. For instance, still limited computer resources do not allow the detailed simulation of the cloud's microphysics on the physical scale of the growth of precipitation drops or nuclei activation. Therefore a bulk water parameterisation technique has to be employed within the prognostic scheme, artificially subdividing the liquid phase into rain water (generated by simple empirical parameterisations for cold and warm clouds) and cloud water (generated if relative humidity exceeds a specific threshold) – as already developed by Edwin Kessler and Hilding Sundqvist (Kessler 1969; Sundqvist 1978; Sundqvist et al. 1989). Such a bulk water parameterisation contains basic processes such as the conversion of cloud water into rain water, the evaporation of rain and the melting of snow. Thereby,

[15] Advection is the process by which some properties of the atmosphere, e.g. heat and humidity, are transported.

the parameterisation of the evaporation of rain 'is uncertain and different schemes can give quite different evaporation rates' (Tiedtke 1993, 3046). Furthermore, bulk water parameterisation is adjusted by using coefficients for important processes, such as the Bergeron-Findeisen process, which have been modelled only recently.[16] Although the prognostic cloud scheme of Tiedtke (1993) required fewer assumptions than the diagnostic schemes used up to that time, it still used various 'disposable parameters' to close the parameterisation, for instance 'a humidity threshold value to initiate the formation of stratiform clouds [...] and a diffusion coefficient for the turbulent mixing of cloud air and environmental air' (Tiedtke 1993, 3060).

8.3.2.1 Tying Together Epistemically Diverse Knowledge

The interesting question is: how are the epistemically diverse ingredients of a cloud scheme composed on the level of code? The view into a current cloud scheme of a climate model – in our case the ECHAM5 model[17] – unveils the modellers' strategy. The stratiform cloud scheme of the ECHAM5 model consists of three conceptual parts: a cloud microphysical scheme, a statistical cloud cover scheme and three prognostic equations for the grid-cell mean mass mixing ratios of water vapour, cloud liquid water and cloud ice. These three prognostic equations are based on 33 terms of sinks and sources of cloud water content, which refer to 23 different sinks and sources. This means that some terms appear as sources in one equation, while they are considered as sinks in the others. For instance, the instantaneous melting of cloud ice (Qmli), if the temperature exceeds the freezing point, is a source for the computation of cloud liquid water, while it is a sink for the computation of cloud ice.

This concept of sources and sinks ties together all terms and each of these terms has to be parameterised, meaning that for each term a scheme has to be formulated explicitly, for example for the instantaneous melting of cloud ice (Qmli). The parameterisation of all 23 different sinks and sources employ a diverse body of knowledge. For example, the stochastical and heterogeneous freezing (Qfrs) – a source for the computation of cloud ice and a sink for the computation of cloud liquid water – is based on an extrapolated equation with constants derived from laboratory experiments. The autoconversion of cloud liquid water (Qaut) – a sink for the computation of cloud ice – is based on a stochastic collection equation, describing the time evolution of a droplet spectrum, which changes through

[16] The Wegener-Bergeron-Findeisen (WBF) process refers to the rapid growth of ice crystals at temperature between 0°C and –35°C, which results from the difference in the saturation of vapour over water compared to that over ice (see the following footnote on the recent modelling of the WBF process).

[17] ECHAM is one of the general circulation models used to compute scenarios for the Assessment Reports of the Intergovernmental Panel on Climate Change (IPCC). The general prognostic variables are vorticity, divergence, humidity, temperature, surface pressure and the mean mixing ratio of water vapour and total cloud water (liquid and ice) (Lohmann and Roeckner 1996).

collisions among droplets of different size. This stochastic collection equation contains a tunable parameter, 'which determines the efficiency of the autoconversion process and, hence, cloud lifetime' (Roekner et al. 2003, 70). Similarly, the accretion of cloud liquid water by rain (Qracl) – a sink for the computation of cloud liquid water – contains a tunable parameter and, furthermore, includes the complete parameterisation of the autoconversion process. And so forth.

These examples show that parameterisation weaves a complex texture of information and scientific knowledge that constitute the subscale processes of a climate model. Within the philosophical framework suggested in the introduction, the example presented is a paradigmatic body of epistemically diverse 'theoretical' knowledge applied in climate models – theoretical in the sense of 'all kinds of scientific knowledge expressed semiotically' beyond the domain of the epistemically uniform body of high-level theory (the adiabatic part). All in all, the stratiform cloud scheme consists of an epistemically diverse body of more than 70 equations, some of which date back to concepts from the 1940s, for example, the Marshall-Palmer distribution (Marshall and Palmer 1948), the extrapolated equation for stochastical and heterogeneous freezing by Bigg (1953), and the mass-weighted fall velocity of rain drops parameterisation according to Kessler (1969).

Table 8.3 Derivation of parameterisation and its epistemic diversity. Source: Gramelsberger and Feichter (2011).

First principles	An analytical solution or an approximation based on some simplifications can be derived.
Laboratory studies	Utilisation of data from laboratory studies because it is too difficult to measure the process in-situ. An example is the study of ice crystal formation in cirrus clouds. The advantage of laboratory studies is that they take place under controlled conditions.
Measurement campaigns	Data from focused measurement campaigns of various continental and marine sites are used to derive robust relationships between various parameters. The information is prepared in compiled datasets, which represent the spatial and temporal variability of the parameterised process. It has to be mentioned that measurement data represent every influence on the investigated process, whether these influences are known or not. In this regard, this method complements the laboratory method for processes that are more complex than can be studied in a laboratory setting. The sample size in a field experiment is normally not large enough to stratify these empirical data according to all influences in question.
Models	Data and information from models with finer resolution are used to derive parameterised processes that occur on small scales. Their statistical behaviour can be described by a stochastic relationship, which is derived from model simulations with finer resolution that are able to resolve some of the processes in question. This method is questionable, as it lacks an observational database.

Other concepts for parameterisation have been developed recently.[18] However, in order to make these concepts computable they have to be translated into statements in the code. This translation can also been seen as the concretisation of each mathematical term and concept employed in the subscale parameterisation.

8.3.2.2 Concretising While Coding

Every part of the stratiform cloud scheme (the three prognostic equations, the cloud microphysical scheme, and the statistical cloud cover scheme) has to be translated into code in order to generate a computable version of the conceived model (see Table 8.1). Therefore, every part of the parameterisation has to be translated into stepwise instructions for computing. For instance, the instantaneous melting of cloud ice (Qmli) when the temperature exceeds the freezing point – a source for the computation of cloud liquid water, while it is a sink for the computation of cloud ice – is coded in nine lines of Fortran code (see the cloud.f90 file, ECHAMS (2005)).

Although Qmli has been conceived simply as the ratio of cloud ice for two time steps ($Qmli = ri/\Delta t$), the computation of the instantaneous melting of cloud ice requires more concrete information. For instance, the gravitational acceleration of 9.80665 ms^{-2} is needed, the amount of heat consumed by evaporation has to be taken into account as well as the heat released by sublimation in order to adequately compute the instantaneous melting of cloud ice. Finally, the amount of cloud water resulting from the melting of snow has to be computed, by adding the available water of the previous time step of computation to the amount of melted snow of the current time step. Additionally, the available cloud ice of the previous time step of computation has to be reduced by the amount of melted snow of the current time step. The coding of the instantaneous melting of cloud ice is based on nine computable statements, and each statement carries out a step of the computation, thereby propagating the simulation model forward in time.

A current atmosphere model like ECHAM5, used for computing scenarios for the Assessment Reports of the Intergovernmental Panel on Climate Change (IPCC), consists of 15,891 declarative and 40,826 executable statements, which contain and apply all of the scientific ingredients employed in the model (ECHAM 2005). Every change to the simulation model adds or deletes statements of the code. Some of the statements are legacy code of several decades of coding, which pass down pieces of code from one model generation to the other; some of the code is brand new. As climate modelling has a tradition going back to the very first weather model by Charney et al. in 1950, these models can be seen as growing organisms incorporating the conjoint knowledge of the entire climate research community.

[18] For instance, the Wegener-Bergeron-Findeisen (WBF) process has been parameterised recently. This process was already described by Alfred Wegener back in 1911 and extended by Tor Bergeron in 1935, whose findings have been confirmed in experimental studies by Walter Findeisen in 1938. Although this process is important, general circulation models account for it in an extremely simplified manner. Therefore, an adequate parameterisation was urgently required and Storelvmo et al. have developed such a parameterisation for the current Oslo-CAM model (Storelvmo et al. 2008).

In the case of the IPCC Assessment Reports there are about twenty of these models involved worldwide – all coded in Fortran. Fortran (Formula Translator), a computer language first introduced in 1954 by John Backus et al., is easy to read and is used as a lingua franca within the climate research community.[19] It facilitates the exchange of lines of code, which can be handed over from one model community to the other. Since the 1950s this practice has generated a family of climate models inheriting parts of each other (Edwards 2000).

8.3.3 Summary – From Ingredients to Statements

The example of cloud parameterisation unveils interesting aspects of the inner world of numerical models. A zoom into the subscale parameterisations leads to various modelling strategies like prescribing, diagnosing or prognosticating. Which strategy is used depends on the empirical knowledge available about the relevant processes, as well as the computer capacities. In general, climate modelling tends to replace prescriptions by diagnostic schemes and, finally, to replace diagnostic by prognostic schemes. This trend can also be described as increasingly introducing a 'sound physical basis'. In order to use a better physical representation of a process, sufficient quantitative knowledge about this process has to be made available by measurement. Therefore, the improvement of subscale parameterisation drives meteorological research.[20]

However, introducing a 'sound physical basis' can also be seen as development towards concretising the model. This is the common strategy for meteorology to improve its models and make them more reliable. As reliability is the core attribute for using models in the policy context, three major ways of achieving a better reliability can be analysed from the given example:

- First, better reliability is achieved by increasing temporal and spatial resolution, enabling representations more suited to the physical scale instead of bulk parameterisations.
- Second, better reliability is achieved by increasingly replacing knowledge derived from local empirical measurements, laboratory experiment and simulation studies with a sound physical basis (theory) to achieve better representations of relevant processes. Thus, prescribing can be transformed into diagnosing, and diagnosing into prognosing.

---

[19] What makes Fortran easy to read is the fact that, according to Backus, its development was based on the question 'what could be done now to ease the programmer's job? Once asked, the answer to this question had to be: Let him use mathematical notations' (Backus 1980, 131). Modellers can easily translate their mathematical models into code and read the code of others.

[20] For instance, the development of better cloud parameterisations is an urgent task for meteorology, coordinated internationally since the 1990s by the Global Cloud System Study (GCSS) project. GCSS is a project of the Global Energy and Water Cycle Experiment (GEWEX), which is one of the core projects of the United Nation's World Climate Research Programme (GCSS 2011).

- Third, better reliability is achieved by the concretisation within the code in order to realise computation. That means that every constant has to be given a concrete value, and that the initial conditions have to be derived from measurement data. Better measurement data therefore ensures better reliability.

Although these levels depend on each other, the lack of computer power for higher resolution and more complex models, as well as the lack of empirical and theoretical knowledge for improving schemes, will always result in diverse and incomplete climate models. On the one hand, these diverse and incomplete models fuel research that continually endeavours to improve these models. On the other hand, this diversity and incompleteness cause major problems in evaluating climate models and this, in turn, yields major challenges for use in policy contexts, as every ingredient has to be tested on its own as well as for its role as a part of the entire model. More ingredients means that more tests are required for evaluation.

8.4 The Outer World Relations of Infectious Disease and Climate – Policy

Reliability in scientific modelling is a measure for the factors that create trust in model-produced results and in the functioning of the model itself. These factors can be varied and, in the case of simulation models, can be categorised into three classes corresponding to the three terms of *validation*, *evaluation* and *verification* (Oreskes et al. 1994). While the term evaluation refers to the investigation of the adequacy of a model for a specific purpose, the concept of validation is related to aspects of mathematics and coding. The model must be consistent and logical, the methods have to be applied correctly, meaning that no coding errors, inadequate numerical methods, or similar faults are allowed. Sometimes the term verification is used, but this term is misleading, as complex systems are impossible to verify – they cannot be proven to be true – because knowledge about complex systems is always incomplete and observational data are not sufficient.

Although policy makers would like to rely on verified models and results, they have to settle for models that have been mathematically validated and scientifically evaluated. In particular, the latter is analysed in the literature about simulation models as tools for supporting policy decisions. The main criterion for the quality of a model is that it agrees to some extent with observations and measurements and that this agreement can be evaluated. This is not as easy to achieve, as it may seem that simulation models create floods of numbers. Whereas in the early days results were compared to measurements analysed by mere visual inspection, and qualified subjectively by statements like good agreement, in the meantime statistical methods have been applied to assess the degree of agreement between model and observation in a more objective way.

Performance metrics have been developed to measure the skill of a model and to establish a set of standard metrics to measure the strengths and weaknesses of a given model, in particular in climate modelling. However, a good agreement with observational data gives feedback only on the quality of the entire model. Model biases, errors and inadequate representations are not easy to detect in this way.

Therefore, the inner world of models has to be analysed and tested ingredient by ingredient. This, of course, is laborious work, but it is this work which fuels model development by improving or replacing ingredients or by expanding the model with new ones.

In the case of climate modelling, parameterisation schemes are tested in isolation and compared to in-situ measurements, to study the behaviour of the scheme. In a further step, the scheme is tested in the framework of the whole system, for example, within an atmosphere model, by switching the new scheme on and off. Finally the full climate system model is integrated and the results are compared to global measurement datasets. Similar evaluation strategies can also be found in infectious disease modelling, for example, for the compartmental structure of a model.

But there is an important aspect to which the literature usually does not refer when evaluation strategies are discussed. This aspect comes to the fore only when a careful view into the inner world of models is undertaken. As the example of cloud parameterisation has shown, reliability is generated during modelling by telling (coding) a 'better story' than which had been told before in the model. A better story is one that has been increasingly concretised, for example by applying a sound physical basis, by increasing the temporal and spatial resolution or by providing a better scenario. In general, storytelling refers to a better narration told by the model. The idea of describing a model in terms of narratives is related to Mary Morgan's case analysis of economic models (Morgan 2001). In her view models are generally seen as narrative devices, which are directed by good questions.[21]

8.4.1 Telling Better Stories

The idea of telling a better story is located on the inner-world level of conceivable ingredients, and takes the individual modeller into account. It is his or her expertise – in accordance with the available literature and data – that ensures the attribute 'better'. Although simulation models to some extent automatise the process of knowledge production, it is the modeller who decides what body of hypotheses, and how, he or she wants to delegate to the machine. Of course, this delegation is not a purely individual choice; it follows the style of a specific thought collective (Fleck 1979).[22] Nevertheless, there are individual preferences rooted in experience and knowledge involved, to which the scientific community refers to by identifying good modellers. For example, the climate modeller Akio Arakawa became famous for his trick of holding kinetic energy constant, which hardly reflects realistic conditions (Küppers & Lenhard 2005).

---

[21] In contrast, the idea of 'storytelling with' code as developed by Gramelsberger (2006, 2010), refers to the numerical concretisation for computing a simulation model.

[22] In the 1930s Ludwik Fleck developed the concept of the thought collective (*Denkkollektiv*), which has become important in philosophy of science. In his case study of the development of a typhus serum, he identified the different thought-styles of various research groups and their ability to support or hinder new developments by being looked at in such thought collectives (Fleck 1979).

However, this trick helped him to generate more realistic results than other climate models at that time, because 'only Arakawa's model had the aperiodic behavior typical of the real atmosphere in extratropical latitudes, and his results were therefore used as a guide to predictability of the real atmosphere' (Phillips 2000, xxix). Arakawa's model told a better story insofar as it avoided some computational problems that had led to illusory wave patterns. Arakawa was trained as a mathematician in a different thought collective, and therefore had different strategies at his disposal, which yielded a different story. As the example of cloud parameterisation has shown, there are various possibilities of telling a story better, either for computational, scientific or methodological reasons.

In the case of infectious disease modelling story-telling has a further notion, as story-telling here regards building predictive scenarios as a process of model manipulations. These predictions are stories of how an infection potentially behaves under particular circumstances. In the case of Hib models (see Sect. 8.2.2), predictive scenarios allowed the researcher to create a 'playground' and experiment by modelling the effects of different vaccination schedules, or of different actions with regard to an infected individual in a group. On the basis of these predictions, a vaccination schedule was optimised for particular age-groups in order to minimise the circulation of Hib bacteria in the population. As story-telling itself implies, these predictive scenarios are, after all, one possible narrative that answers questions like 'how things might develop' or 'what would happen if'. Their usefulness lies in the fact that they encourage exploring alternative scenarios to optimise the policy outcomes in a given situation.

However, the results of a story told by a model are usually floods of data, and these floods have to be translated into policy-relevant arguments. In the case of climate models, the work of the Intergovernmental Panel on Climate Change (IPCC) has helped to translate model results into levels of confidence, or a scale of likelihood for policy makers. For instance, a more than 90% probability of a result is designated very likely, and one beneath 33% unlikely (IPCC 2005; Petersen 2006). In the wording of IPCC 'it is extremely likely that humans have exerted a substantial warming influence on climate' since 1750 (IPCC 2007, 131). Similarly, in health policy a scenario of a widespread epidemic can be evaluated as likely or very likely.

8.5 Conclusion

To sum up, the outer world of simulation models, which is their predictive and prognostic power, is, of course, determined by the concept of their inner world. The way modellers conceive these inner worlds is directed and constrained by computational resources, empirical knowledge and good scenarios, and measurement data. However, the driving factor in climate modelling as well as in infectious disease modelling is to tell better and better stories. The attribute 'better' is clearly defined by the common practices of specific thought collectives. In case of climate modelling 'better' refers to more physical representations of process, which, in turn, means decreasing the epistemic diversity of a model, for example increasingly rooting the model in theory – theory in the narrow sense as outlined in Sect. 8.1. 'Better' in this sense is denied for application-driven models, as the example of infectious disease modelling has shown. Here 'better' refers to more realistic scenarios and reasonable answers to 'what would happen if' questions. For other disciplines there might be different practices and measures.

The way we have addressed the epistemic diversity of models and analysed their inner world has shown how conceivable ingredients and computable statements simultaneously create the reliable core of models, but create restrictions regarding the broader applicability of models to challenges from the outer world. These aspects demand further investigations into the practice of computer-based simulation and constitute a fruitful base for comparative studies across modelling fields.

References

Auranen, K.: On Bayesian Modelling of Recurrent Infections, Rolf Nevanlinna Institute, Faculty of Science. University of Helsinki, Helsinki (1999)

Auranen, K., Ranta, J., et al.: A statistical model of transmission of Hib bacteria in a family. Statistics in Medicine 15(20), 2235–2252 (1996)

Backus, J.: Programming in America in the 1950s –Some Personal Impressions. In: Metropolis, N., Howlett, J., Rotta, G.C. (eds.) A History of Computing in the Twentieth Century, pp. 125–135. Academic Press, New York (1980)

Bigg, E.K.: The supercooling of water. Proceedings of the Royal Society of London B 66(688), 688–694 (1953)

Bjerknes, V.: Das Problem der Wettervorhersage, betrachtet von Standpunkt der Mechanik und Physik. Meteorologische Zeitschrift 21(1), 1–7 (1904)

Boumans, M.: Built-in justification. In: Morgan, M., Morrison, M. (eds.) Models as Mediators, pp. 66–96. Cambridge University Press, Cambridge (1999)

Boumans, M.: The Reliability of an Instrument. Social Epistemology 18(2-3), 215–246 (2004)

Charney, J.G., Fjørtof, J., von Neumann, J.: Numerical Integration of the Barotropic Vorticity Equation. Tellus 2(4), 237–254 (1950)

Dear, P.: Disciplines & Experience. The Mathematical Way in the Scientific Revolution. Chicago University Press, Chicago (1995)

Daley, D., Gani, J.: Epidemic Modelling. An Introduction. Cambridge University Press, Cambridge (1999)

Dowling, D.C.: Experiments on Theories: the construction of scientific computer simulation. University of Melbourne, Melbourne (1998)

ECHAM5, cloud.f90 file. Max Planck Institute for Meteorology, Hamburg (2005)

Fleck, L.: The Genesis and Development of a Scientific Fact. University of Chicago Press, Chicago (1979)

Edwards, P.N.: A Brief History of Atmospheric General Circulation Modeling. In: Randall, D.A. (ed.) General Circulation Model Development, pp. 67–90. Academic Press, San Diego (2000)

GCSS GEWEX Cloud System Study (2011), http://www.gewex.org/gcss.html (accessed June 20, 2011)

Gramelsberger, G.: Story Telling with Code – Archaeology of Climate Modelling. TeamEthno–Online 2, 77–84 (2006), http://www.teamethno-online/Issue2/ (accessed June 20, 2011)

Gramelsberger, G.: Conceiving meteorology as the exact science of the atmosphere – Vilhelm Bjerknes revolutionary paper from 1904. Meteorologische Zeitschrift 18(6), 663–667 (2009)
Gramelsberger, G.: Computerexperimente. Wandel der Wissenschaften im Zeitalter des Computers. Transcript, Bielefeld (2010)
Gramelsberger, G.: What do numerical models really represent? Studies in History and Philosophy of Science 42(2), 296–302 (2011)
Gramelsberger, G., Feichter, J.: Climate Change and Policy – The Calculability of Climate Change and the Challenge of Uncertainty. Springer, Heidelberg (2011)
Habbema, J.D., De Vlas, S.J., et al.: The microsimulation approach to epidemiologic modeling of helminthic infections, with special reference to schistosomiasis. American Journal of Tropical Medicine and Hygiene 55(5), 165–169 (1996)
Hamer, W.H.: Epidemic Disease in England. The Lancet 1, 733–739 (1906)
Hartmann, S.: The World as a Process. In: Hegselmann, R., Müller, U., Troitzsch, K.G. (eds.) Modelling and Simulation in the Social Sciences from the Philosophy of Science Point of View, pp. 77–100. Kluwer Academics Publisher, Dordrecht (1996)
Humphreys, P.: Extending Ourselves. Computational Sciences, Empiricism, and Scientific Method. Oxford University Press, Oxford (2004)
Hurst, C.J., Murphy, P.A.: The transmission and prevention of infectious disease. In: Hurst, C.J. (ed.) Modeling Disease Transmission and its Prevention by Disinfection, pp. 3–54. Cambridge University Press, Cambridge (1996)
IPCC, Guidance Notes for Lead Authors of the IPCC Fourth Assessment Report on Addressing Uncertainties. Intergovernmental Panel on Climate Change, Geneva (2005)
IPCC, Climate Change 2007: The Scientific Basis. Contribution of Working Group I to the Fourth Assessment Report of the Intergovernmental Panel on Climate Change. Cambridge University Press, Cambridge (2007)
Kermack, W.O., McKendrick, A.G.: A Contribution to the Mathematical Theory of Epidemics. Proceedings of the Royal Society of London A 115(772), 700–721 (1927)
Kessler, E.: On the distribution and continuity of water substance on atmospheric circulation. Meteorological Monographs 10(32), 84–102 (1969)
Küppers, G., Lenhard, J.: Computersimulationen: Modellierungen zweiter Ordnung. Journal for General Philosophy of Science 39(2), 305–329 (2005)
Lohmann, U., Roeckner, E.: Design and performance of a new cloud microphysics parameterization developed for the ECHAM4 general circulation model. Climate Dynamics 12(8), 557–572 (1996)
Manabe, S., Smagorinsky, J., Strickler, R.F.: Simulated climatology of a general circulation model with a hydrological cycle. Monthly Weather Review 93(12), 769–798 (1965)
Mansnerus, E.: The lives of facts in mathematical models: a story of population-level disease transmission of Haemophilusinfluenzae type b bacteria. BioSocieties 4(2-3), 207–222 (2009)
Mansnerus, E.: Using models to keep us healthy: Productive Journeys of Facts across Public Health Networks. In: Howlett, P., Morgan, M. (eds.) How Well Do 'Facts' Travel? Dissemination of Reliable Knowledge, pp. 376–402. Cambridge University Press, Cambridge (2010)
Mansnerus, E.: Explanatory and predictive functions of simulation modelling: Case: Haemophilus Influenzae type b dynamic transmission models. In: Gramelsberger, G. (ed.) From Science to Computational Sciences. Studies in the History of Computing and Its Influence on Today's Sciences, pp. 177–194. Diaphanes, Zurich (2011)

Marshall, J.S., Palmer, W.M.: The distribution of raindrops with size. Journal of Meteorology 5, 165–166 (1948)

Mattila, E.: Struggle between specificity and generality: How do infectious disease models become a simulation platform? In: Küppers, G., Lenhard, J., Shinn, T. (eds.) Simulation: Pragmatic Constructions of Reality. Sociology of the Sciences Yearbook, vol. 25, pp. 125–138. Springer, Dordrecht (2006a)

Mattila, E.: Interdisciplinarity 'In the Making': Modelling Infectious Diseases. Perspectives on Science: Historical, Philosophical, Sociological 13(4), 531–553 (2006b)

Mattila, E.: Questions to Artificial Nature: a Philosophical Study of Interdisciplinary Models and their Functions in Scientific Practice. Philosophical Studies from the University of Helsinki, vol. 14. Dark oy, Helsinki (2006c)

Merz, M.: Multiplex and Unfolding: Computer Simulation in Particle Physics. Science in Context 12(2), 293–316 (1999)

Merz, M.: Kontrolle – Widerstand – Ermächtigung: Wie Simulationssoftware Physiker konfiguriert. In: Rammert, W., Schulz-Schaeffer, I. (eds.) Können Maschinen handeln? Soziologische Beiträge zum Verhältnis von Mensch und Technik, pp. 267–290. Campus, Frankfurt (2002)

Morgan, M., Morrison, M.: Models as Mediators. Cambridge University Press, Cambridge (1999)

Morgan, M.: Models, Stories and the Economic World. Journal of Economic Methodology 8(3), 361–384 (2001)

Mäkelä, P.H., Käyhty, H., et al.: Long-term persistence of immunity after immunisation with Haemophilusinfluenzae type b conjugate vaccine. Vaccine 22(2), 287–292 (2003)

Oreskes, N., Shrader-Frechette, K., et al.: Verification, validation and confirmation of numerical models in earth sciences. Science 263(5147), 641–646 (1994)

Petersen, A.C.: Simulating Nature: A Philosophical Study of Computer-Simulation Uncertainties and Their Role in Climate Science and Policy Advice. Het Spinhuis Publishers, Apeldoorn (2006)

Phillips, N.: The Start of Numerical Weather Prediction in the United States. In: Spekat, A. (ed.) 50 Years Numerical Weather Prediction, pp. 13–28. Deutsche Meteorologische Gesellschaft, Berlin (2000)

Roeckner, E., et al.: The Atmospheric General Circulation Model ECHAM5. Model description. Report No. 349. Max Planck Institute for Meteorology, Hamburg (2003), `http://www.mpimet.mpg.de/fileadmin/publikationen/Reports/max_scirep_349.pdf` (accessed June 20, 2011)

Rohrlich, F.: Computer Simulation in the Physical Sciences. In: Fine, A., Frobes, M., Wessels, L. (eds.) Proceedings of the 1990 Biennial Meetings of the Philosophy of Science Association, PSA 1990, pp. 507–518. Philosophy of Science Association, East Lansing (1991)

Simon, H.: The Sciences of the Artificial, 3rd edn. The MIT Press, Cambridge (1996)

Sismondo, S.: Editor´s Introduction: Models, Simulations, and their Objects. In: Sismondo, S., Gissis, G. (eds.) Science in Context, vol. 12(2), pp. 247–260 (1999)

Smagorinsky, J.: General circulation experiments with the primitive equations. I. The basic experiment. Monthly Weather Review 91(3), 99–164 (1963)

Soper, H.E.: The Interpretation of Periodicity in Disease Prevalence. Journal of the Royal Statistical Society 92(1), 34–72 (1929)

Storelvmo, T., Kristjánsson, J.E., et al.: Modeling of the Wegener–Bergeron–Findeisen process – implications for aerosol indirect effects. Environmental Research Letters 3(045001), 1–10 (2008)

Sundqvist, H.: A parameterization scheme for non-convective condensation including prediction of cloud water content. Quarterly Journal of the Royal Meteorological Society 104(441), 677–690 (1978)

Sundqvist, H., Berge, E., Kristjansson, J.E.: Condensation and cloud parameterization studies with a mesoscale numerical weather prediction model. Monthly Weather Review 117(8), 1641–1657 (1989)

Suppes, F.: A Comparison of the Meaning and Uses of Models in Mathematics and the Empirical Sciences. Synthese 12(2/3), 287–301 (1960)

Thiedeke, M.: Representations of Clouds in Large-Scale Models. Monthly Weather Review 121(11), 3040–3061 (1993)

van Fraassen, B.: The Scientific Image. Clarendon Press, Oxford (1980)

Winsberg, E.: Sanctioning Models: The Epistemology of Simulation. Science in Context 12(2), 275–292 (1999)

Winsberg, E.: Simulations, Models and Theories: Complex Physical Systems and Their Representations. Philosophy of Science 68(3), S442–S454 (2001)

Winsberg, E.: Computer Simulation and the Philosophy of Science. Philosophy Compass 4(5), 835–845 (2009)

Woodward, J.: Making Things Happen: a Theory of Causal Explanation. Oxford University Press, Oxford (2003)

Chapter 9
Modelling with Experience: Construal and Construction for Software

Meurig Beynon

University of Warwick, UK

Abstract. Software development presents exceptionally demanding conceptual challenges for model-building. It has such diverse phases, may touch so many disciplines, may address personal and public applications, can involve collaboration of experts from many fields, user participation, and an essential need for ongoing revision. Software modellers must see and think like designers, logicians, engineers, programmers, business analysts, artists, sociologists. This chapter reviews thinking about software development with particular reference to: the limitations of adopting the formal representations that classical computer science commends; different approaches to rationalising the use of models in software development; and the problems of conceptualising software development as theory-building. It concludes by sketching an embryonic approach to software development based on 'Empirical Modelling (EM)' that draws on William James's pluralist philosophy of 'radical empiricism', the historian of science David Gooding's account of the role of 'construals' in experimental science, and the sociologist Bruno Latour's vexing notion of 'construction'. The products of EM development are interactive artefacts whose potential meanings are mediated by the patterns of observables, dependencies and agencies that they embody, as elicited by the actions of the participants in the model-building.

9.1 Introduction

In many disciplines the use of models is widespread and essential. The ubiquity, variety and usefulness of models is undeniable. Sometimes the model is more concrete than what is being modelled (for example, the moving beads of an abacus corresponding to steps in a calculation), sometimes it is less concrete (for example, a differential equation referring to fish stocks), but typically the model is a simplification, with only selected features of the object or phenomenon being modelled. Sometimes the model refers to something 'given' (for example, the solar system), sometimes the referent is as yet only imagined (for example, a building or a design for a building). In each of the examples just offered the model is (or

C. Bissell and C. Dillon (Eds.): Ways of Thinking, Ways of Seeing, ACES 1, pp. 197–228.
springerlink.com

could easily be) a formal representation. At least in the sciences the everyday use of such models has become second nature to practitioners and their value and validity are, for the most part, taken for granted.

Many applications of modelling in science and engineering exploit computing as a way of implementing pre-existing models derived from theory or established empirical knowledge. Creating software in such a context is a routine design exercise. The focus in this chapter is on the challenges encountered in developing software where such methods either do not apply or do not deliver a suitably engaging product – for instance, in complex systems that require more open-ended radical design (cf. Jackson 2005, Vincenti 1993), or in applications in humanities computing where software is developed to provoke the scholar's imagination (cf. McCarty 2005). Software development of this nature, drawing on its strong links with engineering, mathematics and human interaction, makes pervasive but problematic use of a wide assortment of models both informal and formal in character. This chapter draws attention to some of the challenging problems that are highlighted by such software development, with particular reference to the nature of the modelling it entails (cf. Mahr 2009), the semantic range of the models that it invokes, and the need to make formal and informal representations coherent. It then sketches a new approach, Empirical Modelling (EM)[1], in which the modelling processes are based firmly on direct, 'living' experience rather than formal representations. The principal activity in EM is the development of provisional, personal sense-making entities ('construals') that, in an appropriately engineered context, can be interpreted as public entities amenable to formal description.

In order to explain the approach of EM, three fundamental questions concerning modelling have to be addressed:

- How can direct, personal experience serve as a 'grounding' for modelling?
- How can we exploit and justify 'construals' as a means of building models?
- How can the constructed, mediated character of the 'public entities' developed by modelling be reconciled with the 'real-world' and 'formal logical' entities that play such essential roles in conceiving and implementing complex systems?

The three main themes of the chapter are framed by these questions. These themes are not treated here systematically or in depth; they recur and are interwoven as necessary to elaborate our thesis about modelling (in Sect. 9.9). Each theme is explored in some detail by, respectively, the philosopher William James (cf. James 1912), the historian of science David Gooding (cf. 1990, 2007) and the sociologist Bruno Latour (cf. 2003). A deeper appreciation of how these themes relate to each other and to EM can be gained from these references and other EM publications (see Beynon 2005; Beynon & Harfield 2007; Beynon & Russ 2008).

It is widely taken for granted that in order to produce a computer solution, or a programmed solution, a given problem must first be replaced by an abstract version. For example, the physical symbols or pieces and their locations in a game, or the liquid levels in an experiment, are replaced by values in a mathematical, or a programming language, type. Then it is this abstract version that is actually solved

[1] See www.dcs.warwick.ac.uk/modelling. Accessed 2 June 2011.

by the program, or investigated by a model or simulation. Simplistic as this formulation undoubtedly is, the fundamental assumption is pervasive and deep: a computer-based solution primarily engages with an abstract version of a problem or task, in contrast to the problem or task as we experience it. In the latter case – but not in the former – there are typically physical components such as symbols, traffic lights, documents, measurements etc., and with all such things there are borderline cases, unclear or partial cases, mistaken cases etc., arising from human psychology and context.

This chapter introduces an approach to modelling, and thereby an approach to computing (in particular to software construction), developed over many years at the University of Warwick, which reverses the broad direction of the problem-solving process just described. The concrete personal experience of a problem instance is taken here as primary, while abstractions, wherever they are useful, take an auxiliary role. The approach is known as *Empirical* Modelling because of its emphasis on observation and experiment; it promotes an alternative way of using computers: one that gives priority to our direct experience of phenomena. This is made possible by regarding the computer itself (with its associated devices) as a source of new experience that is continuous with, or of the same kind as, experience of the real-world problem or phenomenon.

The EM project has been established for about 25 years. It has generated a number of practical modelling tools and several hundred models/construals. These models have unusual interactive qualities, open-endedness and semantic plasticity. Whereas conventional modelling uses abstraction to simplify complex phenomena, EM generates an interactive environment in which to explore the full range of rich meanings and possible interpretations that surround a specific – typically seemingly 'simple' – phenomenon ('playing a game of noughts-and-crosses' (Beynon & Joy 1994), 'learning to use your digital watch' (Beynon et al 2001), 'explaining a perplexing encounter with a lift' (Beynon 2005), 'solving a Sudoku puzzle' (Beynon & Harfield 2007)). Because of its emphasis on enriching experiences and exposing semantic possibilities, EM can be applied directly in certain areas (such as creative design, education, and the humanities), but in general is best understood as an activity that can complement and enhance software development based on traditional abstraction techniques. Although notions characteristic of EM (such as agent-orientation and dependency) also feature prominently in modern software practice, there is no coherent conceptual framework for their adoption alongside tools and techniques aligned to traditional computational thinking (cf. Roe & Beynon 2007; Pope & Beynon 2010). Making the case for EM as a suitable conceptual framework motivates the fundamental questions concerning modelling raised above.

The chapter considers how the challenges of software development call for the reconciliation of views, and the integration of activities, that are superficially quite divergent in character: formal and informal, personal and public, machine and human based, goal-directed and exploratory. The scene is set (in Sects. 9.2 and 9.3) by highlighting contrasting perspectives that are associated with the mathematical and engineering aspects of software and with the possible roles for the machine and the human within computationalist and constructivist traditions.

Several perspectives on how to bring coherence to the software process are briefly reviewed (Sects 9.4, 9.5 and 9.6); these include high-level critiques of the predominantly 'rationalistic' emphasis endorsed by theoretical computer science (for example, Winograd & Flores (1986); Cantwell-Smith (1987, 2002); Naur (1985, 1995), approaches that propose new methodologies that complement conventional techniques for representing and modelling software (for example, Harel (1988; 2003); Jackson (2000; 2005; 2006; 2008); Kaptelinin and Nardi (2006)) and approaches that tackle the foundational issues surrounding models and their role in development (for example, Mahr 2009; McCarty 2005; Gooding 2007; Addis 1993; and Addis & Gooding 2008). It is in Sect. 9.5 that the key ideas of the chapter emerge as follows. A constructivist account of software calls for a conceptual framework that makes personal experience fundamental and this is supplied in the 'radical empiricism' of William James. Software development demands modelling of both kinds – experiential and formal – but integrating models of such different kinds to afford conceptual coherence has proved a serious challenge.

All the approaches mentioned in the above paragraph operate within a framework in which the software has a semantics conceived as formal (for example, mathematical), though this may be explicitly informed by surrounding informal processes which play a significant part. This is in keeping with an established scientific tradition such as that advocated by James Clerk Maxwell, who maintained that analogy and physical illustration have a crucial role in developing abstract mathematical models. The final sections of the chapter (from Sect. 9.8 onwards) introduce EM as an alternative approach to computing, and to software, that gives priority to immediate experience of the concrete and situated without excluding formal representations, and that invokes ideas drawn from William James for its semantic foundation. EM has much in common with the provisional, personal modelling employed by Faraday in his experimental researches for which Gooding (1990) introduced the term 'making construals'. Developing construals is well-oriented towards the needs of software engineering where addressing human aspects, auditing the development process, and adaptation in response to changing requirements are concerned. It can also be seen as a technique with qualities appropriate for what Latour (2003) conceives as 'construction'. The chapter concludes with a brief account of how EM addresses the problems of modelling for software development mentioned above.

9.2 Ways of Seeing, Ways of Thinking

The development of complex software systems combines abstract mathematical activities relating to formal specification and inference with concrete engineering activities associated with configuring devices and building their interfaces. The range of issues to be addressed in this way is well-represented in Jackson's paper *What can we expect of program verification?* (Jackson 2006), which is strongly rooted in the formal computer science tradition, whilst recognising the need to endorse Vincenti's concern that engineering should 'do justice to the incalculable complexity of the real-world' (Vincenti 1993).

In understanding the different ways of thinking that are represented in mathematics and engineering, it is helpful to distinguish two ways of looking at everyday objects. We can look at an object such as a clock, for instance, with specific attention to those aspects that serve its characteristic function of 'telling the time'. For a digital clock, the relevant observables are starkly symbolic: the digits on the LED display, perhaps with an indicator as to whether it is now 'am' or 'pm'. For an analogue clock, the relevant observables are somewhat more subtle: the hands of the clock and their orientation relative to marks on the face and to each other. In either case, our way of seeing is directed at abstracting specific information about mathematical measurements of time elapsed in hours, minutes and seconds. Contrast the way in which we might observe a natural object such as a tree. Like the clock, a tree maintains its integrity through continuous state change: its growth, its motion in the wind, its response to the changing seasons. But a tree object has no clear function or protocol for observation. There are a whole variety of reasons why we might pay particular attention to different features of the tree, according to whether we are interested in identifying its species, its state of health, its capacity to provide shade or shelter, its potential to cause damage to property and so on. By paying attention in one specific way we do not do justice to all that a particular tree object can be.

A clock and a tree may naturally invite different 'ways of seeing', but this has as much to do with the typical context and disposition of their observer as with their intrinsic nature. A clock is built for a specific purpose and is typically viewed, as it were, through the filter of its function. A tree is typically a given of our natural environment that contributes to our here-and-now experience in ways that are unpredictable and unplanned. Mathematicians enjoy distilling patterns and structures (such as differential equations and groups) through circumscribed ways of observing, interacting and interpreting 'clock-like' objects. Engineers characteristically engage with concrete 'tree-like' objects in more holistic and open-ended ways. But fruitful work in engineering and mathematics draws on both types of observation. In designing an attractive clock, the engineer must take account of all kinds of observations whilst focusing on those that fulfil the narrow requirement. In devising a mathematical model of a tree, the mathematician adopts a constrained way of observing features relevant to a functional objective but may first need to identify suitable features and patterns of behaviour derived from exploratory experiment.

The two 'ways of seeing' we have contrasted are respectively associated with two 'ways of thinking': *functional abstraction* and *exploratory interpretation*. Their co-existence and mutual interplay has been prominent in the history of mathematics and engineering, and has especial relevance in modern computing. On the one hand, digitisation and computational science have made it possible to express ever more complex and subtle systems using functional abstractions. On the other hand, new computing technologies and applications have led to the development of artefacts with unprecedented scope for imaginative exploration and interpretation. The impact of these developments has been to juxtapose two quite different perspectives on computing in a radically new way, potentially setting up a tension that threatens to polarise the discipline. A key concern in this chapter is to sketch an integration and reconceptualisation that can promote both perspectives.

The potential for tension derives in part from the way in which computing now pervades all kinds of disciplines, including many that have well-established modes of enquiry into meaning unlike those of science as traditionally conceived. The accepted theoretical foundation for the digital culture is well-disposed towards rational, logical, objective, reductionist stances. It is well-suited to scientific applications that can exploit a formal account of language and semantics. Applications of computing in an instrumental role in support of human activities in everyday life, business and the arts, by contrast, have more affinity with the notion of language as metaphor and lend themselves to hermeneutic, experiential, subjective and holistic stances. An extreme form of functional abstraction represents the universe as a machine whose behaviour can in principle – and in due course will be – comprehensively observed (Weinberg 2002). It may be contrasted with the extreme variety of exploratory interpretation embraced by the English literary critic I. A. Richards, whose vision of language as metaphor (cf. Lakoff & Johnson 1980) informed:

> [his] compelled and compelling belief in the absolute reality of the world of poetry, which was not, for him, parallel to the world of life but was coextensive with it, and the means of deepest penetration into it. (Vendler, 1981)

If we are to make sense of the way in which computing technology is now being exploited across disciplines we need a conceptual framework in which both of these perspectives can find their place.

9.3 Computationalist and Constructivist Accounts of Agency

The extent to which we can and wish to automate human activities is topical in relation to software development. Software development incorporates sense-making activities carried out in the identification of requirements and goal-oriented rule-based activities that are addressed in implementation. These two kinds of activity are respectively associated with ways of thinking represented by 'exploratory interpretation' and 'functional abstraction'. In conceiving a complex software system, it is necessary to characterise the roles of many agents, human and non-human, external and internal to the system and to engineer their interaction in accordance with their capabilities and the characteristics of the environment. The modelling approach that we advocate is one in which no absolute presumptions about agency need be imposed; at the discretion of the modeller, agents can act and interact autonomously according to law-like rules or be animated under human control. A construal for a system will typically be a hybrid between two extremes: the *computationalist* account of agency in which agent interfaces are based on 'functional abstraction' and the *constructivist* account of agency in which agent interfaces are based on 'exploratory interpretation'. These two kinds of account are introduced below. The benefits of making construals can be fully appreciated only by adopting an agnostic stance about causality (that is, about the 'real' agencies at work) – for which Turing's stance explained below gives some motivation.

In the celebrated paper in which he introduced the so-called 'Turing Test', Turing discussed the issue of whether or not humans were 'regulated by laws of behaviour' and accordingly might be 'some sort of machine' (Turing 1950). At that time, sixty years ago, the idea of proposing that the human mind might be machine-like was perhaps much more implausible than it is today. The very significant impact that computers based on the principles identified by Turing has subsequently had on science in general, and on theories of mind, has encouraged some to promote a computational conceptualisation of the universe as the next step in a sequence of historical developments in science that have displaced perspectives that privilege the human (the earth is not the centre of the universe, nor the sun; humans are biologically of the same nature as animals etc.). Wolfram's 'new kind of science' (Weinberg 2002; Wolfram 2002), Humphreys's vision for 'extending ourselves' (Humphreys 2004), and Bentley and Corne's (2001) aspirations for evolutionary design, invite us to reappraise the roles for the human in making judgements and performing actions. In a similar spirit, some researchers seek to demonstrate that even the core activities of the humanities can be conducted by machines rather than humans (cf. Boden 2003).

For his part, Turing acknowledged that he had 'no very convincing arguments of a positive nature to support [his] views' (Turing 1950) and it is not entirely clear that, even in today's context, he would have endorsed such grand visions for computation. The argument he used in 1950 to undermine the position of those who argued against the possibility of mind-as-machine is still valid: "... we cannot readily convince ourselves of the absence of complete laws of behaviour ... The only way we know of for finding such laws is scientific observation, and we certainly know of no circumstances under which we could say, 'We have searched enough. There are no such laws.'" But when he further comments: 'As I have explained, the problem is mainly one of programming ...', it is not clear that even Turing's visionary imagination could foresee the form that 'the problem of programming' would take by the end of the 20th century. As is widely acknowledged, the problems of complex software systems development, and the limited progress towards resolving them using formal mathematical methods, indicate the need for better ways of taking human aspects into account when exploiting automation.

The problem of finding the proper place for the human in software development can be seen in the broader perspective of controversies surrounding constructivist accounts of science. In 'The Promises of Constructivism', Bruno Latour (2003) spells out a vision for what adopting a constructivist perspective entails. What he declares to be its characteristic conception – that '[e]verywhere, building, creating, constructing, laboring means *to learn how to become sensitive* to the contrary requirements, to the exigencies, to the pressures of conflicting agencies where none of them is really in command' – seems well-matched to the practice of software engineering. What is more, when we consider those issues that have provoked controversy (cf. Latour 2003), a constructivist outlook seems to be much better oriented in some respects towards software engineering than it is towards science. For instance, in analysing the problems faced in promoting a constructivist account of science, Latour writes:

> The two things science studies did *not* need were to replace the fascinating site it was uncovering by an unconstructed, homogeneous, overarching, indisputable 'society' and of course an unconstructed, already there, indisputable 'nature.' This is why science studies found itself fighting on two fronts: the first against critical sociology it wrongly appeared to descend from (as if it were merely extending social explanation coming from law and religion to science and technology); and the second against nature fundamentalists who wanted facts to pop up mysteriously from nowhere. (Latour 2003)

In the case of software engineering, it is entirely appropriate to attribute some aspects of software to laws, norms and power relations within society, and its artificial status is not in doubt. Yet, as Latour explains, there is a much more fundamental concern to be addressed in championing a constructivist viewpoint. Latour is dismayed by the way in which his constructivist vision for science has been discredited by misguided accounts from 'critical sociology' of what *construction* entails:

> ... the problem of construction ... has to do ... with the inner mechanism of construction itself. The problem with constructivism is that no one could account for the building of anything ... by using this metaphor as it has been popularized in social sciences. (Latour 2003)

In his view, rehabilitation of the notion of 'construction' is crucial to redressing the situation. And, by a *good* notion of construction, Latour means a notion of construction set within a conceptual framework that affords scope for debating about – if not discriminating between – 'good' and 'bad' construction.

9.4 Software Development and Constructivism

On the face of it, software development is an obvious topic to consider from a constructivist perspective. Whereas scientists may have reservations about whether scientific facts are already there to be discovered, these reservations do not apply to software engineering products (cf. Latour's discussion in which he contrasts the metaphysical status of a building with that of a scientific fact: '... everyone will agree that ... the building was not there before' (Latour 2003)). Perhaps the principal challenges facing software development are concerned with the activities that elicit requirements and with modifying software in the face of changing requirements. Both of these problems relate directly to those qualities associated with constructed objects – 'history, solidity, multiplicity, uncertainty, heterogeneity, risk taking, fragility, etc.' – that are identified by Latour. And having been built with sensitivity to 'the contrary requirements, to the exigencies, to the pressures of conflicting agencies' is a much more holistic criterion for quality for a software product than 'meeting its functional requirement'.

There are good reasons why traditional software development is not well-aligned to constructivist thinking, however. Software development has been deeply influenced by a way of seeing that is dominated by functional abstraction and a way of thinking that favours formal representations of the kind that Turing's work inspired. These dispositions reflect the core issues that the theory of computing addresses: *how to develop ingenious algorithms to fulfil a given requirement? – how to optimise these to minimise the computation involved? – how to implement them on high-performance architectures? – how to express them in programming languages? – how to analyse them using mathematical logic?* – all of which presume pre-established requirements. Adopting this focus does not prevent us from addressing broader questions. For instance, Mahr's study of models of software (Mahr 2009), which takes as its central question *Does the system S comply with the requirements for its application*, is directed at being able to modify and adapt to requirements in a flexible way, and to address non-functional requirements. But the representations that computer science has developed tend to favour situations in which the requirement has been abstracted and the methods that are to hand engage with these abstractions (cf. Kramer's characterisation of computer science as 'the study of abstraction' (Kramer 2007)).

Many meta-level accounts have challenged the notion that classical computer science concepts can give an adequate account of computing activity, and of software development in particular. Prominent amongst these are the writings of Winograd and Flores (1986) and of Brian Cantwell-Smith (1987; 2002). Naur (1985; 1995) is another fierce critic of the accepted theoretical computer science culture who has championed the idea that intuition has an essential role in software development. A key concern in these critiques is that the semantic relation that links a computer program with its application domain (which Cantwell-Smith (1987) terms 'the semantics of the semantics of the program' in deference to the narrow machine-oriented interpretation of 'the semantics of the program' in computer science) is established and maintained because of human, social and contextual factors that cannot be formalised.

Such critiques highlight limitations of theory without addressing what Latour has identified as the key question for the constructivist: *what form should good construction of software take?* Critiquing the accepted theoretical framework of computer science does not necessarily give insights into a practice that has developed its own independent culture. For instance, although the traditional characterisation of models adopted in software development – 'representations of a system at many levels of abstraction, and from many viewpoints' – might suggest an emphasis on formal representations that characterise the product rather than its construction, such models in practice exploit informal representations to record histories and variants, and to document design decisions. And whilst there may be good empirical evidence for discounting some informal but ephemeral representations that have been advocated for software development in the past, we should be sceptical of the notion that our ability to formalise is necessarily the best measure of our understanding. In this connection, Black's caveat regarding the use of mathematical models in science is pertinent:

> The drastic simplifications demanded of success of the mathematical analysis entail serious risk of confusing accuracy of the mathematics with strength of empirical verification in the original field. (Black 1962)

9.5 Formal Representations and Experientially-Mediated Associations

Many agents have roles in the development of a software system. They may be human and non-human, internal and external to the system. The ways in which these agents actually or metaphorically observe and act are diverse and reflect their particular viewpoints and capabilities. A fundamental challenge in developing the system is making these many viewpoints coherent by identifying some kind of 'objective reality' in which they can all be taken into account. What Winograd and Flores (1986) describe as a 'rationalistic' approach to development in effect tries to use representations suited to specifying objective knowledge to address all the modes of 'knowing' that are represented in the development environment. As Brian Cantwell-Smith remarks, formal representations are insufficiently nuanced to reflect the rich indexical character of the communication associated with development: they encourage 'promiscuous modelling' in which there is no differentiation between models that are merely logically equivalent (cf. every computer is a Turing machine when characterised as an abstract computational device) (Cantwell-Smith 1987). In defending such approaches, it is hard to avoid what Latour terms a 'fundamentalist' position.

Latour characterises *fundamentalism* as 'deny[ing] the constructed and mediated characters of the entities whose public existence has nonetheless to be discussed', and suggests that 'constructivism might be our only defense against fundamentalism' (Latour 2003). Even if we were to subscribe to the idea that all human behaviour could be explained by 'computational' rules, this would not necessarily be good grounds for taking a fundamentalist stance. It is quite apparent for instance that knowledge of the laws of projectile motion is neither necessary nor sufficient to guarantee that a person can intercept a ball in flight. By the same token, knowing that our actions are determined by laws, and being able to correlate our actions with these laws as they are being enacted are two quite different things.

To sustain the idea of constructivism as affirming 'the constructed and mediated character of the entities whose public existence has to be discussed', it is essential to have some means of referring to personal experience. For this purpose, we shall adopt a key tenet of William James's radical empiricism: that knowing something is to experience an association between one aspect of our experience and another (James 1912/1996). We shall refer to such associations as 'experientially-mediated'. To appreciate the highly personal and context-dependent nature of experientially-mediated associations, we might consider the different associations that would be established if a person looking at the first page of the Moonlight Sonata were to be (say) a child who has not learnt to read music *or* a professional pianist who was about to give a performance *or* a future archaeologist unfamiliar with the entire Western musical tradition.

Constructivism is centrally concerned with activities like learning and discovery that link the personal to the public: engineering the context, training the agents, discovering the boundaries so as to build bridges between the two ways of seeing identified above. The roles for formal representations and experientially-mediated associations in these activities are illustrated by considering what is involved in 'learning to read the score of the Moonlight Sonata', or might be involved far in the future in 'rediscovering how to read the score'.

On the one hand, the score of the Moonlight Sonata has the status of a formal representation. To interpret this representation, we have to respect public conventions about how to observe and apply preconceived rules so as to read the score as a functional abstraction. There are also experientially-mediated associations that can be invoked by the score which go far beyond what is expressed in the formal representation: the apocryphal back-story to its composition known to the child; the tempo and the shapes and movements of the hand that are prepared in the pianist's mind; the fragmentary remnants of a piano keyboard dug up by the archaeologist. The difficulty that we face in exploiting these associations effectively is that they are much more subjective in character, more contingent on specific context and have to be mediated in much more complex and obscure ways; consider, for instance, what is demanded of the pianist wishing to communicate her sense of preparedness to play. (Compare the challenges of communicating techniques to the user of a brain-computer interface in contemporary applications of software development.)

In general, practising computer scientists acknowledge that both formal and informal experientially-mediated representations have an essential role to play in software development, and venture to exploit both without concern for how – perhaps even *whether* – they can be coherently integrated *at the conceptual level.* Software development is a prime example of a field in which both kinds of representation come together to create intimidating conceptual challenges. The diversity of the proposals for tackling this issue is an indication of just *how* intimidating. Harel, approaching the field from the perspective of formal software development, recognises the importance of experientially-mediated representations to the developer; he devised the statechart as a 'visual formalism' for this purpose (Harel 1988), and has subsequently sought to exploit message sequence charts as a way of enabling end-users to participate in the formal specification of systems (Harel & Marelly 2003). Jackson (2006), drawing on decades of experience as a leading software consultant, identifies the vital need to engage with the environment in the spirit of the engineer when the development of a complex software system entails non-routine design; he stresses the dangers of premature commitment to specific object-based decompositions and advocates the use of 'problem frames' (Jackson 2000; 2005) as a means of critiquing decompositions. Nardi (1993), who has drawn on her background in anthropology in her extensive studies of software development, highlights the way in which specific representations and social contexts influence the development process; she favours an account of software development that is cast in the framework – similar in spirit to that advocated by Winograd and Flores (1986) – of Activity Theory (Kaptelinin & Nardi 2006).

9.6 In Search of Principles for Modelling Based on Experientially-Mediated Associations

The system development activities discussed by Harel, Jackson and Nardi all rely on the incremental construction of what are traditionally termed 'models' of the system to be delivered. Such a 'model' may be quite amorphous in character, being informed by diverse elements such as UML diagrams, logical specifications, prototype code and informally expressed requirements and documentation. As Loomes and Nehaniv (2001) have pointed out, a major conceptual difficulty is to justify the idea that throughout the development process the evolving models have integrity and can in some way be shown to be authentically associated with the yet-to-be-built final system to which they at all times refer. A key problem is that an abstract functional specification of a system presumes a degree of assured and objective understanding that the developer in general only acquires through the development process. In different ways, Harel, Jackson and Nardi are concerned with how experientially-mediated representations that actively call for human interpretation and judgement can be exploited in reaching a suitable specification. To address Loomes's concern, it is necessary to understand what principles, if any, legitimise the use of such informal representations in these approaches. Specifically, we need to know in what sense, notwithstanding the misconceptions, revisions, and refinements to which the model is subject, the development of a system can be regarded as a coherent and discriminating process of *construction*.

If construction is to demonstrate 'the constructed and mediated character of the entities whose public existence has to be discussed', it must be based on the incremental development of a succession of models. If the construction is addressing a problem requiring 'radical design', the models developed in the initial stages will typically exploit experientially-mediated associations, and only in later stages become integrated with, or identified with, models based on formal representations. To address the key problem of identifying what constitutes 'good' and 'bad' construction (cf. Latour 2003) we need principles by which to assess the quality of human judgements involved in the construction process.

Several different strands of independent research have ventured to provide a foundation for aspects of computing (cf. Beynon et al's (2006) notion of 'human computing') that rely on experientially-mediated associations in some essential way.

Mahr (2009), in his account of modelling in software development, identifies 'a theory of modelling' as crucial to computer science. His observation that 'every object can be a model in some context' motivates a significant conceptual move that shifts the emphasis from 'defining the notion of a model' to 'reasoning about human judgements to the effect that something is a model'. This is the basis for a 'logic of models' in which 'the structure of contextual relationships' has a critical role:

> The logic of models permits an abstract description of the *argument* by means of which a judgement, assumed to be indivisible, and according to which a specific object is assigned the role of being a particular model, may be justified .... the argument ... relies on the existing of [a] structure of contextual relationships as its condition of correctness. (Mahr 2009)

McCarty (2005) identifies modelling as the core activity of Humanities Computing. His key insight is that the semantic force of models in the humanities is in the modelling processes that surround their development and use. Although he chooses to interpret such models within the traditional Turing framework, they derive their expressive power from a dynamic human context in which their meanings are always being interactively negotiated, and where the role of the model is characteristically provocative and problematising.

A third account of how objective knowledge is constructed is that given by Addis and Gooding (cf. Addis 1993; Addis and Gooding 2008; Gooding 2007). This account reflects philosophical ideas from C. S. Peirce and proposes 'a pragmatic stance ... that relates directly our actions in the world to ... our knowledge about the world' (Addis 1993). A key feature is that it draws on what Gooding, in his account of Faraday's experimental work on electromagnetism, calls 'construals'. Such a construal (or tentative model) directly exploits the experientially-mediated associations and 'cannot be grasped independently of the exploratory behaviour that produces it or the ostensive practices whereby an observer tries to convey it.' (Gooding 2007, 88).

Each of these approaches highlights the significance of experientially-mediated associations, and each attempts to find an effective way of connecting them with formal representations: in Mahr's case, reasoning about judgments (Mahr 2009); in McCarty's case, associating meaning with the modelling process rather than model as product (McCarty 2005); in Addis and Gooding's case, transforming empirical knowledge to logical form by a process of abduction (Addis 1993; Addis and Gooding 2008). The vexing question of defining the notion of *model* is prominent in all three treatments. If construction is to be based on a succession of models then it must involve models that have quite different ontological status. This possibility is consistent with Achinstein's conception of:

> ...a hierarchy of models based on their ontological commitments: at one end, we have models which are simply supposed to provide possible mechanisms for how natural systems might be operating, while at the other end, we have concrete claims that the real world is thoroughly like the entities and deductions in our model. In the latter case, the choice of model amounts to a metaphysical commitment regarding the contents of the universe - which things and relations really exist. (Achinstein 1968)

But the versatility of models in this respect gives them a chameleon-like character that has confounded attempts to say precisely what a model *is*. It is hard to devise a definition that takes account of where a model is located along the axes personal/public, subjective/objective, provisional/assured, and the ontological status we may wish to claim for it.

9.7 Software Engineering as Science?

As the previous discussions have shown, notwithstanding the challenges that software development presents to formal approaches, the idea of invoking logical and mathematical models is well-established. There are several reasons for this:

- Many of the most important developments in mainstream computing are based on exploiting pre-existing mathematical and scientific models. In this context, the theory of the application domain is already established, and the role for computer science is to optimise the use of resources through the design and implementation of efficient algorithms and computing environments. The visions of Wolfram (Weinberg 2002; Wolfram 2002), Humphreys (2004) and Bentley and Corne (2001) alluded to earlier are squarely in this tradition, and envisage ways in which theory and computation can have radical transformative impact.
- Theories, which can be mediated symbolically, are well-suited to exploitation on computers as traditionally conceived. Theoretical computer science does not give a good account of applications that cannot be framed in terms of symbolic representations; indeed, it can be seen as calling into question whether such applications are well-conceived. Though the idea that the process of software development follows a 'waterfall' sequence is discredited, there is a presumption that something logically equivalent to a specification has been constructed by the time this development is completed, even if this specification is not made explicit. This presumes that the behaviour of the resulting system is comprehensively captured in terms of protocols for agents, both human and non-human, that are based on functional abstractions.
- A legitimate concern for the theoretical computer scientist is the extent to which a software system without a formal specification can be regarded as a 'public entity'. Clarity about the interpretation of the software by machine and human on the part of all the participating agents seemingly cannot be achieved without recourse to formalism.

A natural consequence of this outlook is that software engineering aspires to be a form of 'theory-building' (such as is envisaged by Turski and Maibaum (1987)). Jackson draws attention to the problematic nature of software engineering as 'science', highlighting the need to do justice to the subtlety and complexity of the engineering concerns that underlie complex software systems (Jackson 2006), and to steer a proper course between formality and informality (Jackson 2000) by exploiting two kinds of model he attributes to Ackoff (1962): 'analytic models that are simplified descriptions of the real world' and analogic models (like databases) that create a 'new reality' to be kept consistent with the real world (Jackson 2008).

An instructive parallel can be drawn with way in which modelling is viewed in relation to scientific theories. By way of illustration, following Kargon (1969), we may consider the different ways in which two schools of scientific thought in the nineteenth century approached the emerging theory of elasticity. The 'mechanico-molecular' school postulated a physical theory based on deriving the properties of bodies from the putative structure and forces governing their molecules. Their aim was to demonstrate that the predictions deduced from this theory were corroborated by experience, and regarded this as confirmation of their hypothesis. The school of 'purely analytical mechanicians' by contrast sought general formulae based on abstract high-level principles that obviated the need for complicated computation of mechanical forces. An 1821 paper by Navier developed equations of motion for elastic bodies from the 'mechanico-molecular' viewpoint; these equations were subsequently derived without reference to any physical theory from a mathematical function describing a elastic body under strain discovered by George Green (Kargon 1969).

In broad terms, these two approaches to describing an elastic body correspond to two complementary ways of specifying a computation to determine its behaviour: in terms of atomic data and rules, or via an equational constraint. Either might be the basis for approaches to computing the behaviour of an elastic body using software, and they are representative of familiar computational paradigms. As explained by Kargon, Maxwell appreciated the motivation for 'simplification and reduction of the results of previous investigation to a form in which the mind can grasp them', but was critical of both approaches. In postulating a physical structure without beginning with the phenomenon to be explained, the molecularists 'are liable to that blindness to facts and rashness in assumption which a partial explanation encourages.' In framing a mathematical formula in the spirit of the analytical mechanicians 'we entirely lose sight of the phenomena to be explained.' In Maxwell's own treatment of elastic bodies (Maxwell 1890), he explained how the plausible physical theory adopted by Navier and others made predictions at odds with experimental observations. He then described an alternative physical theory informed by a dependency relation that could be experimentally verified linking the compression of the solid to the pressure applied at a point, and showed that this produced results more faithful to experiment.

As described by Kargon, concern that abstract mathematical models should be rooted in concrete physical phenomena was central to Maxwell's concept of proper scientific practice. His methods drew upon physical analogies or illustrations such as Faraday developed for electromagnetic phenomena in his experimental researches as well as 'scientific illustrations' that posited common mathematical forms associated with concepts and laws drawn from different branches of science. Such analogies, intended to aid the intuition in the development of theories but then to be discarded, had a role in ensuring that mathematical models were more than a basis for blind calculation. As Maxwell explained in his address to the British Association in 1870, here summarised by Kargon:

> 'The human mind,' he said, 'is seldom satisfied, and is certainly never exercising its highest functions, when it is doing the work of a calculating machine.' The man of science must, of course, sometimes *temporarily* perform the role of such a calculating machine. However, the goal of this labor ultimately is the clarification of his ideas. (Kargon 1969)

Despite the strong motivations for adopting it set out above, the conceptual framework surrounding theory-building in science does not marry well with the challenges of software engineering in key respects:

- The notion of a theory associated with a simplifying mathematical model to clarify our ideas has its place in natural science, where we may be entitled to expect some uniformity and elegance. The same qualities most definitely do not apply to many of the artificial and specific environments and components with which a software system must engage.
- Scientific theories are intended to serve a foundational function and have a degree of permanence that reflects the stability of the empirical world from which

they are derived. It may be appropriate to discard the intuitions that are used to scaffold the construction of such a theory once it is established. The principal challenge in software engineering, by contrast, is to build conceptual structures that typically need to be revised regularly in response to changes to the world in which they operate.
- Software systems have to take account of human agents and personal, social and cultural factors that, at the very least, cannot be readily captured in laws or studied through experiment.

In what follows, we shall sketch a radical solution to the impasse that software engineering seems to have reached through having to reconcile formal representations with the complex, dynamic and human characteristics of the world of experience. The central constituents are: a different, and broader, notion of computer-based modelling ('Empirical Modelling') that involves a radical generalisation of basic principles that underlie the spreadsheet and exploits the power of computing technology as a source of interactive experience; a semantic account in the spirit of James that favours experientially-mediated associations as the basis of 'knowing'; and a pragmatic stance to objective, or negotiated, meaning that is in line with Latour's aspirations for *construction*.

9.8 An Account of Models Based on Experientially-Mediated Associations

The key factor that distinguishes a model that relies on experientially-mediated associations from one based on formal representations is the essential role for personal judgment in the conception of the model. This is in keeping with Wartofsky's (1979) characterisation of a model (cited by Lloyd (2001)) as a *triadic* relation: '*a person* takes *something* as a model of *something else*' and is central to the treatments of modelling by Mahr (2009), McCarty (2005) and Gooding (1990). For model-building to serve as a process of *construction* the model itself and the contextual relationships surrounding it, being an artefact that testifies to 'the constructed and mediated character of a public entity', must evolve in such a way that its integrity is respected. The complexity of this process of evolution is reflected in the many axes along which the model migrates in the modeller's understanding: from provisional to assured; subjective to objective; specific to generic; personal to public.

To support this migration the model has to be fashioned, via interactions both initiated and automated, so that the views of many agents – human and non-human – are taken into consideration. This process of construction is in general open-ended and never absolutely resolved. These features manifest in different ways according to the applications in mind. In software engineering, as Mahr observes, the essential need for open-endedness stems from the way in which the implementation of a system affects its environment. In McCarty's vision for humanities computing, modelling is the means to support and enrich the negotiation of meanings with no expectation of ultimate closure. In his account of Faraday's experimental activities, Gooding traces the way in which construals are revised and developed in conjunction with new conceptions and realisations.

Empirical Modelling (EM) is a body of principles and tools for developing models that are based on experientially-mediated associations. In EM the role for the computer is broadly parallel to the role it plays in a spreadsheet. Consider, for example, a spreadsheet of examination marks. In the first instance, the computer maintains dependencies, so that when a value in the spreadsheet (for example, the mark awarded to Alice for biology) is changed, other values that are contingent upon it (such as Alice's overall average mark) are updated automatically. The nature of these dependencies is determined by the context and intended interpretation. The spreadsheet fulfills its semantic function if the user is readily able to connect each of its cells with meaningful external observables, and if the ways in which changes to these observables are synchronised on interaction respect these meanings (such as when Alice's mark is increased, her overall average also increases).

The meaning of a formal representation is expressed in propositional statements specifying relationships between observables that pertain universally. In the experientially-mediated associations of the spreadsheet, meaning is expressed via latent relationships between observables that reflect 'expectations about potential immediate change in the current state'. Such a representation is highly nuanced in precisely those ways that are most relevant to model-building in a constructivist idiom. 'Potential for change', 'expectations' and 'immediacy' are implicit references to agency and context. In interactions with a spreadsheet prototype that embrace exploratory design and development, the meanings of the spreadsheet – as mediated by different agents acting and observing in different roles – are associated with patterns of interaction and interpretation with which the model-builder becomes familiar over time.

Whereas formal representations are suited to expressing objective information about systems, experientially-mediated associations are suited to tracing what might be termed 'personal' views of 'live' entities. A formal representation draws on previous experience, with a view to abstracting insights to sum up this experience by eliciting the implicit constraints and rules that are deemed to govern change. Associations that are experientially mediated are oriented towards capturing the impressions we experience moment by moment and recording the way in which these impressions cumulatively and dynamically build up conceptions of entities. Where the use of the term 'experience' in relation to formal representations is qualified by 'previous' to convey the idea of reflection and rationalisation, the reference in experientially-mediated associations is to current living experience, in the spirit of Dewey's 'actual focusing of the world at one point in a focus of immediate shining apparency …' (Dewey 1916, ch.1). And where the characteristic emphasis in formal representations is on abstract rules for interpreting symbols, the semantic connections that are invoked by experientially-mediated associations are themselves 'given in experience' as endorsed by William James's *radical empiricist* stance.

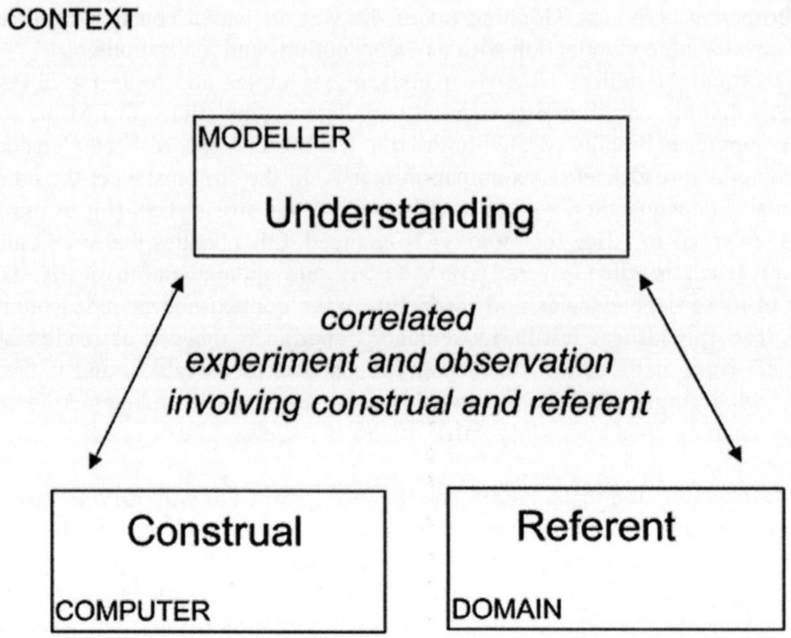

Fig. 9.1 The key ingredients of Empirical Modelling.

In EM, three live ingredients of the modelling environment co-evolve: the computer-based construal, the external phenomenon to which the construal refers the modeller's understanding, and the context within which interaction with the construal is being interpreted. The relationship between these ingredients is represented diagrammatically in Fig. 9.1; superficially an unremarkable diagram that might be drawn to express the core relationships associated with any variety of modelling, but one whose interpretation here is unusual since these relationships are themselves live.

In appreciating this, it is helpful to consider how the corresponding ingredients of a spreadsheet are experienced. In deploying a spreadsheet, the values and definitions that will be associated with observables represented by cells will depend upon the current context ('When will the chemistry results be available?'), relate to agency over which we have limited control ('What mark will Alice get in chemistry?'), are subject to be modified in ways that were not preconceived ('The marks for Bill and Ben Smith were entered the wrong way round') and that might involve exceptional kinds of agency ('It would be useful to add a column to record how many first class marks candidates get in core subjects'), and prompt changes to the modeller's understanding ('I thought that Bill and Ben's marks were out of line with their overall performance'). Equally diverse are the ways in which the spreadsheet can be viewed: 'How did Bill perform overall?'; 'How did students perform in my module?'; 'How transparent is the computation of each student's overall mark?'; 'How does the average mark for my module compare with that of other modules?'. In this respect, the range of possible observational viewpoints and affordances is well-matched to what is required of construction, as discussed above.

The fundamental EM concepts of an *observable*, an *agency* and a *dependency* that have here been illustrated with reference to the spreadsheet are the basis for a very general variety of modelling that can be applied across many disciplines. The tabular interface that mediates state in a spreadsheet is not representative of general EM construals. The net of observables and dependencies associated with a situation is rich and open-ended, and a variety of different modes of visualisation and manipulation is appropriate if the perspectives and 'experience' of the diverse agents who observe and interact are to be reflected. It might be appropriate to attach a line-drawing visualisation to the examination mark spreadsheet, for instance, so that the seating arrangements for an examination could be displayed, or to establish a dependency between a timeline and the spreadsheet to indicate which examinations took place in a particular period. The limits on the possible scenarios and extensions to a construal that can be investigated in this way are set only by the modeller's imagination and the quality of the supporting tools.

The scope of EM is best appreciated by considering the many different kinds of agency, human and non-human, internal and external, that are associated with a complex system that combines human, software and hardware components. Relevant agents in such a setting include systems designers, software developers, users and managers, agents automated in software, devices such as sensors and actuators, and environmental sources of state change such as weather, noise and light. The ontological status of the components in an EM construal differs according to the application, as illustrated in the EM project archive[2]. A construal may have as its referent: a 'real-world' scenario in which the capabilities of individual agencies are well-understood but their corporate behaviour is to be investigated (for example, as in a railway accident); a radical design activity in which some of the agents are pre-existing devices but others are in the process of being developed or programmed in an exploratory fashion; a personal experience (such as appreciating a piece of music, or working in a restaurant). The most important distinction between an EM construal and a conventional model is that a construal is a source of live interactive experience for the modeller, who is free to rehearse interactions with it and explore possible interpretations. In this respect, it is radically different in character from a model that is conceived as a simplification of its unfathomably rich real-world referent (cf. a typical 'scientific model') and is derived by a process of abstraction. A parallel may be drawn with the distinction between writing prose and writing poetry to which the poet Don Paterson (2004) alludes. Paterson argues that ambiguity is an essential quality of poetic use of language, in sharp contrast to uses of language that aim at precision and clarity of expression.

---

[2] The Empirical Modelling project archive, University of Warwick, UK. http:/empublic.dcs.warwick.uk/projects. Accessed 13 June 2011.

9.9 Empirical Modelling as *Construction*

The problem of understanding the nature of modelling required in software development can be seen as linked to the problem of identifying an acceptable concept of *construction* to which Latour (2003) alludes. To give a good account of how 'the entities whose public existence has to be discussed' are constructed, it is essential to have a way of 'representing' what is *not yet* a public entity that can gain the approval of critics from the whole spectrum of viewpoints identified by Latour. The use of the term 'representing' in this context is slightly paradoxical, as 'representing' itself has overtones of 'making public'. But it is quite apparent that, in order to convince the "scientific" sceptic, we must be able to defend and give evidence to substantiate a claim that one entity is associated with another by invoking some device other than formal representation. And – in order to confound the deconstructionist – we must be able to demonstrate that these associations are not arbitrary products of the imagination that can have no compelling public authority.

Our thesis is that Empirical Modelling principles provide just such a means of creating associations as is required of a legitimate notion of construction. The crucial difference between the experientially-mediated associations established in EM and formal representations is that they are, of their essence, contextualised; they relate to the live experience of human observers in the presence of a customised environment for exploratory interaction. To some degree the plausibility of this thesis can be assessed simply by seeking empirical corroboration, but it also commends – perhaps even demands – a philosophical reorientation of just such a kind as is the central theme of William James's *Essays in Radical Empiricism* (James 1912/1996). Specifically, our thesis only makes sense subject to accepting certain prerequisite ideas:

- It presumes experientially-mediated associations can be empirically given in the experience of a human modeller. The force of 'empirically given' is that an association has no further explanation: an association either is or is not experienced. Both these assumptions are in line with James's radical empiricist stance.
- It deems such associations to be objectively real provided that interaction with other modellers indicates that they too experience the 'same' associations. The nature of the interaction involved in justifying the claim of 'shared experience' is discussed at length by James. In keeping with James's stance, his account explains how this interaction is itself rooted in experientially-mediated associations.
- It recognises a role for interactive artefacts in communication that goes beyond the use of formal representations, and in particular, mediates understanding with reference to the way in which change of state of an interactive artefact can be effected and experienced by the modeller. In this way, the most primitive knowledge takes the form of expectations latent in the current state to the effect that 'if this feature changes, it will conceptually in one and the same transition affect other features in a certain recognisable way'. These expectations (or

'dependencies' as they are described in EM) can also be viewed as experientially-mediated associations on account of the way (although they are associated with latent actions) they affect our perception of the current state. The idea that experientially-mediated associations are more expressive than formal representations is (in effect!) also endorsed by James.

The full development of these ideas has been the subject of extensive study relating to EM[3]. The large body of EM construals that have been developed, principally by computer science students who have studied EM at the University of Warwick over the last twenty years, provides some of the empirical corroboration for our thesis. Particularly significant is the range of subjects that have been addressed (construals drawn from engineering, business, science, medicine, music) and the character of the most successful applications, which relate to themes in which the transition from private to public understanding is central (such as learning, requirements cultivation, decision support, personal sense-making).

In the context of this chapter, one of the most relevant themes that has been considered from an EM perspective (cf. Beynon and Russ 2008) is that of 'supporting exploratory experiment'. To illustrate how more useful links between modelling in science and modelling in software engineering might be established, it is helpful to look more closely at the way in which Faraday's experimental researches were conducted, and how their findings were related to the theory of electromagnetism. The adoption of the term 'construal' (rather than 'model') in EM was itself motivated by the way in which Gooding applied this concept in his account of Faraday's work (Gooding 1990). Our speculative proposal is that Faraday's research activities can also be interpreted 'as if' within the conceptual framework of EM, with particular reference to the thesis about constructing public entities set out above.

The development of electromagnetic theory has been a particularly fruitful source of ideas about the significance of models. This is in part because of the wide range of activities represented in this development, from Ampère to Faraday to Maxwell, and the extent to which these have been documented, for example through Maxwell's own reflections (Black 1962; Maxwell 1890) and in the work of Gooding (1990).

At a meeting entitled *Thinking Through Computing*, organised by the EM research group at the University of Warwick, UK, in November 2007, Gooding (2007) reflected on how his notion of 'construal' had influenced the development of EM. He likened the traditional theoretical account of computer science, with its emphasis on characterising behaviours declaratively using functional abstractions, to the mathematical theory that was developed to account for Ampère's studies of electricity. Characteristic of Ampère's researches was the study of electrical phenomena in equilibrium. This was in strong contrast to Faraday's research, in which intervention and dynamic phenomena were key ingredients, for which accordingly pre-existing theory was inadequate. Observing the strong parallel between Faraday's construals and EM artefacts, Gooding remarked that theoretical computer science might presently be in something like the same state as electromagnetic

[3] See www.dcs.warwick.ac.uk/modelling. Accessed 13 June 2011.

theory prior to the work of Maxwell and Thomson. He speculated that, in due course, similar developments in theoretical computer science might accommodate the idea of EM construals through a broadening of the models within its mathematical framework.

Where modelling to support software engineering is concerned (especially in the context of radical design or humanities computing in the spirit of McCarty (2005)), our previous discussion suggests another more radical possibility: that the primary focus of interest is shifted to construals, as complementing mathematical models in an essential way, and as better matched to the demands of the application. The spirit of Gooding's account of Faraday's work is well-expressed in the following quotation from Gorman:

> Gooding believes that Faraday progressed via 'a convergence of successive material arrangements (the apparatus) and successive construals (or tentative models) of the manipulation of the apparatus, and its outcomes' (Gooding 1990, 187). A construal corresponds roughly to what I am calling a mental model. Gooding's point is that, for Faraday, ideas, physical manipulations and objects are closely linked. (Gorman 1997)

The blurring of mind-body duality that is represented here is itself an indication that the construal has neither the idealised qualities of an abstract mathematical model nor the status of a natural reality that a scientist might attribute to its referent. The resistance to accepting such a construal as a useful variety of *model* is vividly reflected in Duhem's (1954) scornful comment on the lamentable judgment of English physicists, as cited by Black:

> In place of a family of ideal lines, conceivable only by reason, he will have a bundle of elastic strings, visible and tangible, firmly glued at both ends to the surfaces of the two conductors, and, when stretched, trying both to connect and to expand. When the two conductors approach each other, he sees the elastic strings drawing close together; then he sees each of them bunch up and grow large. Such is the famous model of electro-static action designed by Faraday and admired as a work of genius by Maxwell and the whole English school. (Black 1962)

Such criticism reflects certain preconceptions about the nature and purposes of 'modelling'. What is being modelled (in this case, 'electro-static action') is being conceived as already part of a public reality, and the model (as defined by mathematical equations) is being understood as a simplification or idealisation derived by abstracting from this pre-existing reality. It is precisely because modelling principles and tools have traditionally been conceived in this way that they are so ill-suited for applications such as exploratory experiment and radical design.

To put Duhem's criticism of Faraday's construals in perspective, it is appropriate to view Faraday's experimental work (albeit on electromagnetic rather than electro-static phenomena) within a constructivist frame of reference as giving insight into 'the constructed and mediated character of the entities whose public existence has to be discussed'. Whereas today an electromagnetic device is 'an entity whose public existence has to be discussed' – most educated people have some notion of how an electric light or an electric motor works – no such public entities were established in the initial stages of Faraday's research.

It seems plausible that Faraday conceived what Gooding characterises as his 'construals' as modelling yet to be discovered causal laws. This would be in accordance with Cantor's account of Faraday's Sandemanian religious faith:

> ... he accepted that nature was governed by a set of God-given causal laws. These laws were ... the work of a wise Creator who, avoiding unnecessary complexity, constructed the world on simple, plain principles. (Cantor 1986)

The preliminary role that these construals served was to represent patterns of interaction and interpretation associated with as yet ill-understood phenomena that began to emerge over an extended period of exploratory experimentation. Identifying such patterns meant establishing contexts in which to discern within the phenomenon itself specific entities whose status might change, means to effect changes to these entities and instances of contingent changes that could be interpreted as a consequence of causal laws.

In order to identify such patterns, and potentially to communicate them to other scientists, it was essential for Faraday to find a means to represent them. The construal served this purpose as a physical construction made of familiar entities (cf. 'a bundle of elastic strings, visible and tangible, firmly glued at both ends to the surfaces of the two conductors' in Faraday's construal of electro-static effects, as described by Duhem) that responded to change on the same pattern, 'as if' obeying the same causal law. As is well-documented in Gooding's account, some of the patterns that Faraday recorded in this way could be reliably reproduced and were embodied in construals, but were nonetheless later deemed to be associated with unsatisfactory experimental techniques or with misconceptions.

The limitations of such construals were not to do with the quality of the experientially-mediated association between an artefact made of commonplace components and a putative 'electromagnetic phenomenon' *per se*, but with their relationship to Faraday's broader aspirations for causal laws that might be deemed to be 'God-given' and 'avoiding unnecessary complexity'. The development of new construals and experimental apparatus and protocols was then aimed at eliminating procedural complexity that required exceptional observational skill and capability on the part of the experimenter, and dependence on features of the local environment. Developments of this kind are precisely aligned with the idea of natural science as relating to universal experience, and with the 'creation' or 'discovery' of 'entities whose public existence has to be discussed'.

As Gooding (1990) observes, what Faraday communicated through his construals was not propositional knowledge; it was a body of interactions in the world, carried out in suitably engineered contexts through exercising certain skills, that followed a predictable and reproducible pattern. On the one hand, this body of interactions could be enacted (either physically, or as a thought-experiment) on the construal as an interactive object whose parts are constituents of our everyday human 'shared experience'. On the other hand, it had a counterpart that could be enacted in the realm of unfamiliar electromagnetic phenomena. Most crucially, the 'same' experientially-mediated association between these two modes of interaction that Faraday engineered personally and locally could also be experienced publicly and universally.

The respect that 'the English physicists' had for Faraday's construals stemmed from their appreciation of the connection they established between the personal and public arena: guiding the intuitions of the learner; connecting abstract mathematical models in mind with interventions in the laboratory; and exposing the deceptive simplicity of after-the-fact accounts of experimental discoveries. As has been explained, making construals of this nature seems much more relevant to the challenges of software engineering, and more realistic, than aspiring to generate 'a mathematical model'. As for the potential relevance of EM in this connection, the above sketchy account of Faraday's approach has been mischievously written in such a way that it refers in all but name to 'observables' ('specific entities whose status might change'), 'agencies' ('means to effect changes to these entities') and 'dependencies' ('instances of contingent changes'). To what extent this might misrepresent Faraday's experimental activities is a question beyond the scope of this chapter that merits further work for which Gooding's studies provide an excellent resource (cf. the computational model of Faraday's cognitive processes proposed by Addis, Gooding and Townsend (1991)).

9.10 Construals and Construction for Software

In conventional software development, the need for what is described by Fetzer (1999) as 'conceptualisation' of a domain prior to the development of use-cases, requirements and specification is well-recognised. The unusual nature of the modelling activity in EM, and its strong affinity with sense-making activities, make it possible in principle to address this conceptualisation in an alternative, constructivist, manner. The parallel between EM construals and Faraday's construals is again instructive. As described in detail by Gooding (cf. Gooding 1990; Gorman 1997), Faraday's construction of a prototype electric motor was the result of a methodical activity through which he was able to infer how forces associated with electromagnetic phenomena could be harnessed by configuring magnets and coils. In EM terms, Faraday's construals can be viewed as embodying insights about the nature of the observables, agencies and dependencies that are characteristic of electromagnetic phenomena. Their dual physical and mental nature enabled him both to experiment and to reason about how to dispose these observables, agencies and dependencies so as to achieve a desired behaviour.

As Gooding (2007) suggests, the characteristics of knowledge representation in traditional software engineering, which focuses on identifying structures, constraints and invariants that are intended not to be subject to change, resemble the mathematical theory developed by Ampère in connection with his research on electromagnetism 'based on experiments in which nothing happens'. Since the 'busy phenomenology' that surrounds construals is a rich resource for the practical engineer, it is unsurprising that Faraday's research generated many more engineering products than Ampère's; building software with EM construals has a somewhat similar power to provoke experimental design and stimulate the imagination.

The potential for developing software based on EM construals has been illustrated in several practical studies (see Beynon et al 2000; Keer et al 2010) and discussed in several previous publications (Beynon et al 1998; 2000; 2006; 2008,

Beynon & Russ 2008). Work is in progress on tools to exploit EM principles in supporting software development on a larger scale (cf. Pope & Beynon 2010), but previous research has established proof of concept for many aspects of the activity. Whereas traditional software engineering has distinct phases (requirements, specification, implementation, testing) in which entirely different activities and representations are featured, development based on construals is much more homogeneous. Just as Faraday's prototype electric motor is 'simply' a way of disposing mechanisms that were embodied in pre-existing construals, so a functional piece of software can be composed from suitably engineered EM construals. The development process is associated with a seamless elaboration of patterns of interaction, engineering of contexts, acquisition of skills in observation and manipulation during which the products of the EM activity undergo conceptual transitions (cf. Fig. 9.1).

In the initial stages, the development of a construal is primarily concerned with identifying the most primitive observables and dependencies in the domain and finding ways to express these using suitable metaphors. At this stage, the emphasis is on making an *artefact* that the modeller can learn to interact with to change observables and recognise dependencies in a predictable way (cf. devising a new scientific or musical instrument). Once such an artefact is suitably refined, so that the modeller can interact conveniently to generate reliable responses, it can be used in a more focused manner to develop counterparts of specific ensembles of observables and dependencies in the domain that can be reliably identified. At this stage the modeller becomes familiar with particular ways of interacting with the EM artefact and its counterpart in the domain, acquires the necessary skills to support these interactions, and learns how to shape the context appropriately. At this point, the artefact comes to serve the sense-making role of a *construal*. Characteristic of interactions at this stage is the open-ended exploratory character of the investigation; the modeller engages in sense-making activities because of uncertainty about the nature of the referent and its possible responses. In a software development context, such activities are appropriate for radical design.

A construal may be legitimately regarded as a *model* when its referent has, or acquires, the status of an established feature or component of the domain and all the characteristic interactions and operations associated with the referent are faithfully reflected in the construal. In a routine design exercise, such components may be known in advance, and devising construals to serve as models of them is a natural initial target for the EM development activity. A notable feature of the development of the prototype electric motor is the way in which state-changes that were first identified through what Gooding (2007) calls 'experiments in which things move, twist and jump' are ' – with some difficulty – constrained in a dynamic, unstable equilibrium'. Software development can be envisaged in an analogous way as, in general 'with some difficulty', engineering contexts in which to place construals so that they autonomously and reliably change state in appropriate ways. Within such a context, a construal has even more constrained behaviour than a model, and may be deemed to be a *program*.

It is difficult to appreciate the character of these transitions as abstractly described; practical experience of an EM exercise is ideally required. It is helpful to keep in mind the idea that an EM artefact/construal/model/program is always presented as a net of observables and dependencies that, like a spreadsheet, is to be interpreted as a live state of affairs. Its classification changes only because the human interpreter wraps it in a different context and conceives it in relation to a different family of pre-engineered interactions and interpretations. By way of illustration (as described in detail in Beynon et al (2000)), an artefact that resembles a heap data structure can *first* be developed into a construal that makes sense of the basic operations associated with sorting using heapsort (such as might be used in teaching), *then* to a model of the heap data structure on which the heapsorting algorithm can be rehearsed manually, *then* to a program in which the steps of the algorithm are fully automated and visually displayed.

The transitions that the EM artefact undergoes in this activity are such that what the modeller can initially interact with in a very unconstrained manner (for instance, as a child might play with a set of jointed rods representing a tree data structure) eventually becomes the focus of a much more disciplined interaction where the interpretation is no longer freely invoked by the modeller's imagination but is prescribed by public conventions about the nature of heapsort. Significantly, the invariants of the algorithm, which are traditionally the basis for its formal specification, here enter as sophisticated kinds of observable that are present so long as the interaction with the EM artefact is constrained to follow the correct protocol.

9.11 EM and Some Key Issues in Modelling for Software

The above discussion highlights three key issues that are topical in modelling for software applications: the nature of models; the semantic scope that their use affords; and the need to combine formal and the informal representations coherently. In conclusion, we consider the potential impact of EM upon these issues.

EM offers a new perspective on the nature of models. Reflection on EM practice reveals the exceptional semantic subtlety of what can be embodied in an artefact by appealing to experientially-mediated associations. Appreciating this subtlety, even as it applies to such a simple algorithmic activity as heapsort, makes it quite apparent why it is so difficult to define the notion of 'model'. Although the term 'transition' has been used to describe the way in which the modeller's conception of an EM artefact can evolve, it is the modeller's conception, as defined by the attendant contexts and patterns of interaction and interpretation, that changes. The artefact itself remains no more than a net of observables and dependencies to which agencies, of which the modeller is but one, can be attached.

Formal language is predicated on being able to associate a specific object or category with a word, but this is not well-matched to the way in which an EM artefact is experienced. In EM activities, object boundaries are blurred; what are initially conceived as distinct nets of observables and dependencies relating to different contexts may be composed (cf. the examination mark grid, the examination seating arrangement) and become a new net of observables and dependencies. Categories are likewise blurred: it may be that, according to the modeller's current purpose and state-of-mind, a given net of observables and dependencies can be as of now taken as artefact, construal, model or program.

The transitions that the artefact undergoes in its relationship to the modeller help to clarify the way in which philosophers of science interpret the term 'model'. Within the frame of reference of science, the transitions can be thought of as typically moving from the personal to the public world, and it is with models that refer to public entities that such philosophers are concerned. When Duhem (1954, 70) speaks disparagingly of the models of the English physicists, and Braithwaite (1953) declares that 'the price of the employment of models is eternal vigilance', the notion of model and the modelling context they have in mind is far from the construals that Gooding characterises as a 'tentative models'. It is not even clear that it is appropriate to interpret Achinstein's reference to 'models which are simply supposed to provide possible mechanisms for how natural systems might be operating' (Achinstein 1968) as conceived as applying to such construals, though that interpretation suits our purpose in this chapter (cf. Sect. 9.6).

More ambiguous – and provocative – is the way in which Black (1962) introduces the term 'conceptual archetype' to refer to the kind of mathematical models akin to the 'scientific illustrations' favoured by Maxwell (of which the analogies between electromagnetic fields and fluid flow is the archetype), and then goes on to liken the use of such conceptual archetypes to 'speculative instruments' in the sense of the English literary scholar I.A. Richards (1955). This draws attention to the second key issue that is relevant to our central theme.

The study of the 'artefact-construal-model-program' axis brings together two quite different cultures, according to whether we put the emphasis on the personal or the public. Taking full account of the personal realm is the aspiration of the arts and humanities, which are concerned with the universal human experience. By contrast, the universality that science seeks relates to the public sphere, and concerns what can be communicated objectively. As is more fully discussed in Beynon et al (2006), the scope for shifting the perspective from personal to public in model-building that EM affords is potentially a basis for bridging the two cultures.

The way in which Black likens Maxwell's 'scientific illustrations' to I.A. Richards's 'speculative instruments' also hints at this possibility. At the end of his chapter, Black writes:

> When the understanding of scientific models and archetypes comes to be regarded as a reputable part of scientific culture, the gap between the sciences and the humanities will have been partly filled. (Black 1962)

The problem of addressing the subject of modelling in its full generality from a scientific perspective has been highlighted in the process of framing the vocabulary for this chapter. (An interesting comparison – and contrast – may be made with the problem discussed by Stephen Wolfram in his blog *The Poetry of Function Naming* (Wolfram 2010), where words to describe a very precise functional abstraction are being sought.) The distinctions between two ways of seeing, between the computationalist and constructivist, between formal representations and experientially-mediated representations, are not sharp distinctions such as science might endorse; they relate to qualitatively different but inseparable aspects of experience that are encountered at one and the same time. The number in the spreadsheet cell *is* an arithmetic abstraction *and* 'Alice's mark for Geometry' *and* 'the reason why Alice is awarded a prize'. Our skill as an examiner is in being able to experience all these associations at once.

The notion of such, sometimes seemingly paradoxical, conjunctions in experience is at the core of James's radical empiricism (James 1912/1996) and is reflected in the ambiguous personal/public qualities of entities under construction (Latour 2003). More extensive and detailed discussion of the affinities between EM and radical empiricism and EM and constructivism, aimed at establishing EM as a 'reputable' activity, can be found in other papers (Beynon 2005; Beynon & Harfield 2007). But, despite the impressive precedent for the use of construals in Faraday's experimental work, these affinities and the evidence of the 'science wars' suggest that it may be some time before EM is 'regarded as a reputable part of *scientific culture*'.

It is perhaps easier to relate the promise of EM to the arts and humanities. As explained by M.M. Lewis (1956) in a review of I.A. Richards's book *Speculative Instruments* (Richards 1955), the problem that motivated Richards to conceive language as a speculative instrument was that 'the exploration of comprehension is the task of devising a system of instruments for comparing meanings' (Richards 1955). And since Richards perceived language as the chief instrument with which we think, he concluded that 'the very instruments we use if we try to say anything non-trivial about language embody in themselves the very problems we hope to use them to explore'. As Faraday's experimental work illustrates so potently, artefacts can have power above and beyond language alone. The same principle is illustrated by considering the way in which different constructions on the 'artefact-construal-model-program' axis can carry nuances of meaning for which none of these words is adequate; nuances which indeed defy lexical definition (cf. the illustrations of a similar nature concerning words such as 'time', 'state' and 'agent' that are discussed in Beynon (1999)). This suggests that it is more appropriate to liken construals in science, rather than conceptual archetypes, to speculative instruments in the humanities.

The blending of meanings that construals afford places the issue of dealing in a coherent fashion with formal and experiential models in a new light. As the heapsort modelling exercise outlined above illustrates, it is possible to situate a formal representation within a context moulded from experientially-mediated associations. Such a formal representation comes into being only because the engineering of the context for interaction and interpretation, and the discretion the modeller exercises in interacting and interpreting, disclose a stable pattern in experience that can be appreciated 'universally'. The acts of 'engineering' and 'exercising discretion' are evidence of a simplification of experience similar in nature to the 'simplification and reduction of the results of previous investigation to a form in which the mind can grasp them' to which Maxwell alludes (as discussed in Sect. 9.7 above). The virtue of modelling with construals is that the modeller can take account of such simplifications without needing to make absolute commitments; the patterns of observables, dependencies and agency that sustain different kinds of interaction and interpretation are always fluid and open to revision, and can be organised and recorded in such a way that the modeller can switch viewpoint at will.

As mentioned by Muldoon (2006, 18), Faraday prefaced the first edition of his first book *Chemical Manipulation* (1827) with Trevoux's slogan 'Ce n'est pas assez de savoir les principes, il faut savoir MANIPULER' – 'It is not enough to know the principles, it is necessary to know how to MANIPULATE'. For Faraday, construals were a means of representing and communicating ideas richer than formal mathematical models. According to the thesis set out in Sect. 9.9 above, the expressive power of construals stems from their capacity to embody patterns of observables, dependencies and agency. In Faraday's time, the scope for this embodiment was limited by the available technology; many of the interactions with construals that Faraday conceived were enacted only in his imagination. It is in part on account of these technological limitations that such importance is attached to developing theoretical models that can to some extent obviate the need for 'manipulation' by distilling abstract 'principles'.

Building software on the basis of functional abstractions is likewise a technique that was developed with the limitations of the early computing technology in mind; it permits that high degree of optimisation that remains one of the central preoccupations of computer science. But the pervasive influence of these two simplifying strategies should not seduce the computer scientist into imagining that all experience can be conceived in terms of functional abstractions. As the modes of knowledge representation for AI advocated by Brooks (1991a; 1991b) indicate, the theory of computing must also embrace principles that take account of what it means to 'know how to manipulate'. This chapter has set out to show why EM is a promising basis for such a theory. It also hints at what might be achieved by investing as much effort in developing effective tools for developing construals as has been dedicated to tools for developing programs.

Acknowledgments. I am much indebted to Steve Russ for drawing my attention to useful resources, for supplying the title and most of the introduction, for invaluable editorial help and for vital contributions he has made to the development of key ideas. I also wish to thank Russell Boyatt, Nicolas Pope and many other students who have studied EM for sharing their insights into software development.

References

Achinstein, P.: Concepts of Science: A Philosophical Analysis. The Johns Hopkins Press, Baltimore (1968)

Ackoff, R.L.: Scientific Method: Optimising Applied Research Decisions. Wiley, New York (1962)

Addis, T.R.: Knowledge Science: a Pragmatic Approach to Research in Expert Systems. In: Proceedings of the British Computer Society Specialist Group on Expert Systems: Expert Systems 1993, December 13-15, pp. 321–339. St. John's College, Cambridge (1993)

Addis, T.R., Gooding, D.C.: Simulation Methods for an Abductive System in Science. Foundations of Science 13(1), 37–52 (2008)

Addis, T.R., Gooding, D.C., Townsend, J.J.: Modelling Faraday's Discovery of the Electric Motor: An investigation of the application of a functional database language. In: Proceedings of the Fifth European Knowledge Acquisition for Knowledge-Based Systems Workshop, Crieff Hydro, Scotland, May 20-24 (1991)

Bentley, P.J., Corne, D.W. (eds.): Creative Evolutionary Design Systems. Morgan Kaufmann, San Francisco (2001)

Beynon, W.M.: Empirical Modelling and the Foundations of Artificial Intelligence. In: Nehaniv, C.L. (ed.) CMAA 1998. LNCS (LNAI), vol. 1562, pp. 322–364. Springer, Heidelberg (1999)

Beynon, W.M.: Radical Empiricism, Empirical Modelling and the Nature of Knowing. Cognitive Technologies and the Pragmatics of Cognition: Special Issue of Pragmatics and Cognition 13(3), 615–646 (2005)

Beynon, M., Harfield, A.: Lifelong Learning, Empirical Modelling and the Promises of Constructivism. Journal of Computers 2(3), 43–55 (2007)

Beynon, W.M., Joy, M.S.: Computer Programming for Noughts-and-Crosses: New Frontiers. In: Proceedings of the Psychology of Programming Interest Group Conference, pp. 27–37. Open University (January 1994)

Beynon, M., Russ, S.: Experimenting with Computing. Journal of Applied Logic 6(4), 476–489 (2008)

Beynon, W.M., Sun, P.-H.: Empirical Modelling: a New Approach to Understanding Requirements. In: Proceedings of the Eleventh International Conference on Software Engineering and its Applications, Paris, vol. 3 (December 1998)

Beynon, W.M., Rungrattanaubol, J., Sinclair, J.: Formal Specification from an Observation-Oriented Perspective. Journal of Universal Computer Science 6(4), 407–421 (2000)

Beynon, W.M., Ward, A., Maad, S., Wong, A., Rasmequan, S., Russ, S.: The Temposcope: a Computer Instrument for the Idealist Timetabler. In: Proceedings of the Third International Conference on the Practice and Theory of Automated Timetabling, Konstanz, Germany, August 16-18, pp. 153–175 (2000)

Beynon, W.M., Roe, C., Ward, A., Wong, A.: Interactive Situation Models for Cognitive Aspects of User-Artefact Interaction. In: Beynon, M., Nehaniv, C.L., Dautenhahn, K. (eds.) CT 2001. LNCS (LNAI), vol. 2117, pp. 356–372. Springer, Heidelberg (2001)

Beynon, W.M., Boyatt, R.C., Russ, S.B.: Rethinking Programming. In: Proceedings IEEE Third International Conference on Information Technology: New Generations, Las Vegas, April 10-12, pp. 149–154 (2006)

Beynon, W.M., Russ, S.B., McCarty, W.: Human Computing: Modelling with Meaning. Literary and Linguistic Computing 21(2), 141–157 (2006)

Beynon, M., Boyatt, R., Chan, Z.E.: Intuition in Software Development Revisited. In: Proceedings of Twentieth Annual Psychology of Programming Interest Group Conference. Lancaster University, UK (2008)

Black, M.: Models and Archetypes. In: Black, M. (ed.) Models and Metaphors, pp. 219–243. Cornell University Press, Ithaca (1962)

Braithwaite, R.B.: Scientific Explanation. Cambridge University Press, Cambridge (1953)

Boden, M.A.: The Creative Mind: Myths and Mechanisms. Routledge, London (2003)

Brooks, R.A.: Intelligence without Representation. Artificial Intelligence 47(1/3), 139–159 (1991a)

Brooks, R.A.: Intelligence without Reason. In: Proceedings of the International Joint Conference on Artificial Intelligence, pp. 569–595. Morgan Kaufmann, San Mateo (1991b)

Cantor, G.N.: Reading the Book of Nature: The Relation Between Faraday's Religion and His Science. In: Gooding, D., James, F.A.J.L. (eds.) Faraday Rediscovered. Essays on the Life and Work of Michael Faraday 1791-1867. Macmillan, London (1986)

Cantwell-Smith, B.: Two Lessons of Logic. Computational Intelligence 3(1), 214–218 (1987)

Cantwell-Smith, B.: The Foundations of Computing. In: Scheutz, M. (ed.) Computationalism: New Directions, pp. 23–58. MIT Press, Cambridge (2002)

Dewey, J.: Essays in Experimental Logic. University of Chicago, Chicago (1916)

Duhem, P.: The Aim and Structure of Physical Theory. Atheneum, New York (1954)

Fetzer, J.H.: The Role of Models in Computer Science. The Monist 82(1), 20–36 (1999)

Gooding, D.: Experiment and the Making of Meaning: Human Agency in Scientific Observation and Experiment. Kluwer, Dordrecht (1990)

Gorman, M.E.: Discovery as Invention: Michael Faraday. In: Invention and Discovery: A Cognitive Quest (1997), http://cti.itc.virginia.edu/~meg3c/classes/tcc313_inuse/Resources/gorman.html (accessed June 13, 2011)

Harel, D.: On Visual Formalisms. Communications of the Association for Computing Machinery 31(5), 514–530 (1988)

Harel, D., Marelly, T.: Come, Let's Play: Scenario-Based Programming Using LSCs and the Play-Engine. Springer, Heidelberg (2003)

Humphreys, P.: Extending Ourselves: Computational Science, Empiricism, and Scientific Method. Oxford University Press, Oxford (2004)

Jackson, M.A.: Problem Frames: Analysing and Structuring Software Development Problems. Addison-Wesley, New York (2000)

Jackson, M.A.: Problem Frames and Software Engineering. Information and Software Technology 47(14), 903–912 (2005), http://mcs.open.ac.uk/mj665/PFrame10.pdf (accessed June 13, 2011)

Jackson, M.A.: What can we expect from program verification? IEEE Computer 39(10), 53–59 (2006)

Jackson, M.: Automated software engineering: supporting understanding. Automated Software Engineering 15(3/4), 275–281 (2008)

James, W.: Essays in Radical Empiricism. Bison Books, London (1912/1916); (Reprinted from the original 1912 edition by Longmans, Green and Co., New York)

Kaptelinin, V., Nardi, B.A.: Acting with Technology: Activity Theory and Interaction Design. The MIT Press, Cambridge (2006)

Kargon, R.: Model and Analogy in Victorian Science: Maxwell's Critique of the French Physicists. Journal of the History of Ideas 30(3), 423–436 (1969)

Keer, D., Russ, S., Beynon, M.: Computing for construal: an exploratory study of desert ant navigation. Procedia Computer Science 1(1), 2207–2216 (2010)

Kramer, J.: Is abstraction the key to computing? Communications of the Association for Computing Machinery 50(4), 37–42 (2007)

Lakoff, G., Johnson, M.: Metaphors We Live By. University of Chicago Press, Chicago (1980)

Latour, B.: The Promises of Constructivism. In: Ihde, D., Selinger, E. (eds.) Chasing Technoscience: Matrix for Materiality, pp. 27–46. Indiana University Press, Bloomington (2003)

Lewis, M.M.: Review of Speculative Instruments by I.A. Richards. British Journal of Educational Studies 4(2), 177–180 (1956)

Lloyd, E.A.: Models. In: Craig, E. (ed.) Routledge Encyclopedia of Philosophy. Routledge, London (2001)

Loomes, M.J., Nehaniv, C.L.: Fact and Artifact: Reification and Drift in the History and Growth of Interactive Software Systems. In: Beynon, M., Nehaniv, C.L., Dautenhahn, K. (eds.) CT 2001. LNCS (LNAI), vol. 2117, pp. 25–39. Springer, Heidelberg (2001)

Mahr, B.: Information science and the logic of models. Software and System Modeling 8(3), 365–383 (2009)
Maxwell, J.C.: The Scientific Papers of James Clerk Maxwell. In: Niven, W.D. (ed.), vol. 1. Cambridge University Press, Cambridge (1890)
McCarty, W.: Humanities Computing. Palgrave Macmillan, Basingstoke (2005)
Muldoon, C.A.: Shall I Compare Thee To A Pressure Wave? Visualisation, Analogy, Insight and Communication in Physics. PhD thesis, Department of Psychology. University of Bath (May 2006)
Nardi, B.A.: A Small Matter of Programming. The MIT Press, Cambridge (1993)
Naur, P.: Intuition in Software Development. In: Proceedings of the International Conference on Theory and Practice of Software Development (TAPSOFT), vol. 2, pp. 60–79 (1985)
Naur, P.: Knowing and the Mystique of Logic and Rules. Kluwer Academic Publishers, Dordrecht (1995)
Paterson, D.: Rhyme and Reason, T S Eliot Lecture 2004; published in abridged form in The Guardian Review, pp. 34–35 (November 6, 2004), `http://www.poetrylibrary.org.uk/news/poetryscene/?id=20` (accessed May 11, 2011)
Pope, N., Beynon, M.: Empirical Modelling as an unconventional approach to software development. In: Proceedings of SPLASH 2010, Workshop on Flexible Modeling Tools, Reno/Tahoe, Nevada (October 2010)
Richards, I.A.: Speculative Instruments. Routledge and Kegan Paul, London (1955)
Roe, C., Beynon, M.: Dependency by definition in Imagine-d Logo: applications and implications. In: Kalaš, I. (ed.) Proceedings. of the Eleventh European Logo Conference, Bratislava, Slovakia, August 19-24 (2007)
Turing, A.M.: Computing Machinery and Intelligence. Mind 59(236), 433–460 (1950)
Turski, W.M., Maibaum, T.S.E.: The Specification of Computer Programs. Addison-Wesley, New York (1987)
Vendler, H.: I.A. Richards at Harvard Boston Review (April 1981), `http://bostonreview.net/BR06.2/vendler.html` (accessed June 13, 2011)
Vincenti, W.C.: What Engineers Know and How They Know It: Analytical Studies from Aeronautical History. The Johns Hopkins University Press, Baltimore (1993)
Wartofsky, M.W.: Models: Representation and the Scientific Understanding. Kluwer Academic Publishers, Dordrecht (1979)
Weinberg, S.: Is the Universe a Computer? New York Review of Books 49(16) (October 24, 2002), `http://www.nybooks.com/articles/archives/2002/oct/24/is-the-universe-a-computer/` (accessed June 13, 2011)
Winograd, T., Flores, F.: Understanding Computers and Cognition: A New Foundation for Design. Addison-Wesley, New York (1986)
Wolfram, S.: A New Kind of Science. Wolfram Media (2002)
Wolfram, S.: The Poetry of Function Naming (2010), `http://blog.stephenwolfram.com/2010/10/the-poetry-of-function-naming/`(accessed June 13, 2011)

Author Index

Index